T0258934

Köpriyet: Republican Heritage Bridges of Turkey

Köpriyet: Republican Heritage Bridges of Turkey deals with bridges and the construction industry of Turkey during the foundation of the Republic between 1923 and 1940. This book provides a brief summary of the bridge history of the country, but the main focus is on the Early Republic Era. During this period, the bridge-building technique was reborn in a country undergoing a radical transformation. Turkey changed its ruling, secularized and changed its alphabet.

In addition to detailed descriptions of bridges built during this period and of leading innovative engineers, this book provides a first documented overview of heritage bridges in Turkey, focusing on applied techniques known worldwide. Many bridges are documented for the first time in this book, and most of them are heritage bridges and provide significant value in terms of bridge-building technology and application of worldwide techniques. In the waning days of the Ottoman Empire, engineering projects in Turkey were often carried out through "privileges" by foreign companies. The technical personnel were also financed with foreign resources. With the new Republic, Turkey experienced a renaissance in many areas, including technology. This transition from technology import to development and use of local resources is described in detail.

Köpriyet: Republican Heritage Bridges of Turkey provides a wealth of information and documentation on bridges in Turkey from an important era, and aims at those interested in bridge structures and structural engineering history.

Hulya Sonmez Schaap was born in Turkey, completed her masters at the Middle East Technical University and then worked in that country for 15 years as an engineer. She then migrated to Australia to work there for eight years and then moved to Qatar for work. She has always worked as a bridge engineer, designing, reviewing and strengthening bridges. She is a fellow member of Engineering Australia for which she has been active as secretary of the Qatar Chapter of Engineers Australia through volunteering. Since 1998, she has taken an interest in heritage bridges in Turkey and has visited, photographed and investigated more than 150 of them in Turkey. Her interest has since extended globally, and she has visited more than 50 significant bridges worldwide. She publishes her work and findings about bridges in the following blogs: kantaratlas.blogspot.com and kopriyet.blogspot.com for bridges in Turkey and soon kuranyakuna.blogspot.com for Australia.

Köpriyet: Republican Heritage Bridges of Turkey

Hulya Sonmez Schaap

CRC Press
Taylor & Francis Group
Boca Raton   London   New York   Leiden

CRC Press is an imprint of the
Taylor & Francis Group, an **informa** business

A BALKEMA BOOK

Cover image: Hulya Sonmez Schaap

First published 2023
by CRC Press/Balkema
Schipholweg 107C, 2316 XC Leiden, The Netherlands
e-mail: enquiries@taylorandfrancis.com
www.routledge.com – www.taylorandfrancis.com

CRC Press/Balkema is an imprint of the Taylor & Francis Group, an informa business

Library of Congress Cataloging-in-Publication Data
Names: Schaap, Hulya Sonmez, author.
Title: Köpriyet : republican heritage bridges of Turkey / Hulya Sonmez
Schaap.
Description: First edition. | Boca Raton : CRC Press, 2023. | Includes
bibliographical references and index.
Identifiers: LCCN 2022008806 (print) | LCCN 2022008807 (ebook) |
ISBN 9781032007106 (hardback) | ISBN 9781032007144 (paperback) |
ISBN 9781003175278 (ebook)
Subjects: LCSH: Bridges—Turkey—Design and
construction—History—20th
century. | Historic bridges—Turkey.
Classification: LCC TG111 S33 2023 (print) | LCC TG111 (ebook) | DDC
624.209561—dc23/eng/20220624
LC record available at https://lccn.loc.gov/2022008806
LC ebook record available at https://lccn.loc.gov/2022008807

ISBN: 978-1-032-00710-6 (hbk)
ISBN: 978-1-032-00714-4 (pbk)
ISBN: 978-1-003-17527-8 (ebk)

DOI: 10.1201/9781003175278

Typeset in Times New Roman
by codeMantra

Contents

Foreword

This book will take you to the mystic world of bridge builders and their bridges in the rebirth of a nation between the 1920s and the 1960s. In this era of new revolutions, not only the life of citizens is modernized but also the abandoned and rural roads have been transformed into new ones satisfying the comfort level of the new era. Some of these bridges are still in service and have survived major earthquakes. Please also remember that at those times engineers did not have the luxury of the powerful computation techniques nor material supply of our modern age. Such lack of knowledge and tools has always forced the engineers of the past to be creative while utilizing the basic rules of physics. In that sense, this book will enlighten the readers to the construction abilities of that era as well as mysteries in the construction of these magnificent structures.

In this fascinating book, you can also find some introductory information on the bridges of ancient and historical times in Anatolia, built by many great engineers and architects such as Leonardo De Vinci's proposal for the Golden Horn crossing. Also, discover bridges on the İstanbul-Baghdad Line, also known as the famous Orient Express that has been an excellent setting for many movies.

I have enjoyed reading this book on bridges of the new republic and have realized that the conditions were very much limited to build any bridge due to decades of negligence and ongoing endless wars under the previous rule. Under Atatürk's lead, the country has also improved roads and bridges of modern Turkey. On the tenth anniversary of the republic, even a song was written with some important lyrics referring to railways.

I am very glad to read this book in many ways and give thanks to Hülya Sönmez for her diligence and good will in researching, even searching among the dusty shelves of many remote libraries in Turkey with the hope of finding small pieces of information on any bridge construction. I am also very happy that she has successfully put together so many of the scattered pieces of the mysteries of puzzles of that era in one complete book. I strongly suggest that anyone with an interest in bridges, history or Turkey read this breathtaking book.

With all my best

Prof. Dr. Alp Caner
PhD, PE Bridge Engineer
Prof. METU Civil Eng, Ankara Turkey
President of IABSE Foundation, Zurich, Switzerland

Preface

It all started with a book: I was a Junior Engineer working in bridge design when I visited the bookstore of the General Directorate of Highways (KGM) in Turkey. There I discovered an old paperback book lying on the ground covered with a thick layer of dust on it. There was a masonry bridge photograph on its cover page. This book was "Stone Bridges in Anatolia" by Fügen Ilter. The attendant explained that "only a few remain from the last print and after they finished the lot, the book will not be printed again". This seemed a poor fate for a book, and I thought something should be done about it.

Since then, in 1998, I have been working on the topic by visiting heritage bridges and recording their details. My interest began with the stone bridges in Turkey and then evolved and spread to all kinds of bridges anywhere in the world.

My first visit was a holiday trip to Amasya with a friend, who was an amateur photographer. I looked at all the beautiful bridges and came back without taking any photographs. Luckily, my friend shared her photographs with me later.

My second trip was to Bursa, where I would later design many bridges as an Engineer. The bridge, called Nilüfer, was described to be about 20 minutes from the main road to Mudanya; therefore, I and a colleague looked for this bridge for hours in a very far distance than its place. Then I realized the date of the book was the 1970s and the road layouts were changed and the distances described with time were probably based on the use of horse-powered carts, not a car distance to bridges.

Since then, I have improved my photography and history knowledge. My studies evolved as I was visiting and collecting information, documentation, material and everything about these bridges. I visited over 150 bridges in Turkey. I started sharing my studies in 2013 in the blog called Kantaratlas, which stands for "Map of Bridges" in a combination of languages. This blog introduces the heritage historical bridges in Turkey, and describes their history, main characteristics and engineering features. The blog aimed to share information and to promote the care and passion of bridges to those who find something from themselves in these wonderful monuments.

During my visits, bowstring concrete bridges, which were most of the time nearby, charmed me with their robust size, very intrinsic features and extraordinary appearance. They were too outstanding not to be part of the scenery, while they were completely blended with the environment and contributing to it. Then I realized the importance and impressive context of these bridges, which were from the Early Republican Era of 1923 onwards. This was the period when the traditional masonry construction was replaced with modern concrete and steel bridges in Turkey.

In addition, they belong to a very particular period, where a new nation and new country were rebuilding from its ashes. The Republic also meant independence, freedom and modernity. The people constructing these bridges were a generation born in the empire, who lost their homeland, regained it and built their new home. Therefore, the bridges themselves, their builders and the conditions are all unique, and each one of them is a witness to their period.

I thought that these bridges were not many in number and would be relatively quick to cover. I intended to prioritize them, opening a substudy within my ongoing bridge works, and finish them.

My initial plan succeeded partially. I started a separate blog in 2015 called Köpriyet: made up of the word Köprü and Cumhuriyet that would stand for the "Bridges of Republic". I published 75+ bridges but the study kept growing.

During my studies, I realized that these bridges, as a product of a new republic, and the period during which they were built with its conditions, realities and achievements were left under a blanket. This proud and respectful period that every Turkish citizen owes their identity to, has been ignored and remained in the shadow of the glorified light often shone upon the period of the Ottoman Empire. It is essential to understand the Ottomans as our ancestors, but our modern roots are of the Republican tree. The Republic, and its secular and civilized state, is the reason I have written this book with respect and gratefulness for the opportunities provided to its people.

Therefore, this book of the engineering facts, features and stories of these bridges was written to enlighten the Early Republican Period, its conditions, challenges and people through the bridges.

Hülya Sönmez Schaap
Türkiye 2020-Melbourne 2021

Acknowledgements

I am deeply grateful to my husband, Ronald Schaap, for providing all sorts of support and dedication for my studies. We visited many bridges together, and he read the whole book many times with utmost care and diligence.

My brother, Ahmet Sönmez, accompanied me with many bridge visits. My sisters Canan, Zehra and Derya have always supported my studies and provided logistical services in acquiring photographs and books in Turkey and escorting me on my bridge visits during my holidays. My niece, Petek Bilge, helped in translations from French.

My friend and colleague, Prof. Alp Caner, convinced me that this book should be written and supported this study in all phases. He appreciates and always supported my passion on bridges.

Altok Kurşun, a passionate bridge engineer, helped me to contact STFA for photographs and answered all my questions with diligence.

Nur Taşkent and Sezai Taşkent, the third generation of family directors of STFA, were very supportive and kindly opened their company archive for this book. Gülay Göktürk helped me in obtaining the photographs.

Esra Tümen Dinçkök, President of FATEV Foundation, kindly allowed to use of their archives.

Many of my colleagues contributed to this book in reading and commented on the chapters:

Deniz Günal Dinçkal read all introduction chapters thoroughly and provided corrections and comments for the text.

Uluç Ergin, Senior Civil Engineer, read the introduction chapters.

Prof. Dr. Alp Caner read all technical chapters and gave comments.

Prof. Dr. Mehmet Utku commented on technical chapters and provided valuable feedbacks.

Marco Vargas Correa, Principal Structural Engineer, read all technical chapters and provided valuable comments and corrections.

Melike Çınar, Structural Engineer, commented on Chapter 06-Metal Bridges.

Many individuals supported this book by sharing their photographs.

Efkan Sinan provided the photographs for Cendere Bridge, Polio Aqueduct Bridge, and Nif Bridge.

Bob Cortright kindly provided the photographs for Aksu Bridge.

Ertuğrul Ortaç shared photographs for Big Agonya and Tabakhane bridges.

Mehmet Emin Yılmaz shared photographs for the Güreyman (Kız) Bridge.

Prof. Durmuş Öztürk kindly provided fascinating photographs of many bridges and Alikaya and Suçatı bridges used in this book.

Yusuf Köleli helped me to locate bridges in Kahramanmaraş and took the photographs of Tekir Bridge.

My friend, Architect and Photographer Sevil Işıl Ören, shared her photograph of Çayırhan Bridge and took photographs of Nervi Bridge for this book.

Ahmet Sayılır kindly shared photographs of Genç Bridge.

For the translations from Ottoman to Turkish, I received help from a friend, Alkan Tarım, and Teacher Zeynep Akkır.

I am grateful to all individuals who contributed and shared their enthusiasm on the subject.

AUTHOR

Hülya Sönmez Schaap is a Principal Bridge Engineer with more than 25 years of experience in bridge design, assessment, strengthening and review.

She was born in Turkey, where she received her B.Sc. degree in 1994 and M.Sc. Degree in structural dynamics in 1998 from Middle East Technical University. She worked as a design engineer mostly for bridges in Ankara, Turkey, for 12 years. After migrating to Australia in 2006 for a job offer with Sinclair Knight Merz (SKM), she worked in there for 8 years and then went to Qatar working for the Public Works Authority as a review and proof engineer.

She is currently working for Jacobs, Australia.

She is a fellow member of Engineers Australia for which I have been a secretary of the Qatar Chapter.

Since 2000, she has been interested in heritage bridges in Turkey and visited, photographed, and investigated more than 200 bridges heritage bridges in Turkey. Her interest has since extended globally visiting many significant bridges worldwide.

Her studies are published in blogs: kantaratlas.blogspot.com and kopriyet.blogspot.com for bridges in Turkey and soon kuranyakuna.blogspot.com for bridges in Australia.

PHOTO CREDITS

The captions of the photograph include the credit line of the photograph sources in a shortened form. Therefore, the source details will be expanded here, in the order of appearance in this book:

Photographs by the author are taken during the site visit of the author and their dates are from 2000 to 2021; dates are excluded from captions.

SALT Archive:

SALT is a cultural institution in public service producing research-based exhibitions, publications, web and digitization projects, as well as developing programmes including screenings, conferences and workshops, founded by Garanti BBVA in 2011.

Website: Saltonline.org

BOA and BCA:
   BOA: Başbakanlık Osmanli Arşivi (Archives of the Ottoman)
   BCA: Başbakanlık Cumhuriyet Arşivi (Archives of the Republic)
   Website: https://www.devletarsivleri.gov.tr/

İBB Atatürk Library
   İstanbul Metropolitan Municipality Taksim Atatürk Library
   Atatürk Library is the initial library of early republic period.
   I would like to thank Metin Tekden and Zeki Arslan for their kind help during the photograph selections and providing permissions.
   Website: http://ataturkkitapligi.ibb.gov.tr

FATEV Archive: Feyzi Akkaya Temel Eğitim Vakfı
   Feyzi Akkaya Basic Education Foundation was founded on 24 November 1978 by Prominent Engineer Feyzi Akkaya. He donated nearly all his assets to this foundation and also the copyright of his book "Ömrümüzün Kilometre Taşları (1989)", which was referenced frequently in this book.
   Photographs are mostly taken by Feyzi Akkaya during his visits as Nafia engineer. http://www.fatev.com/

STFA Archive:
   Company established by two prominent engineers of early Republic engineers, Sezai Türkeş and Feyzi Akkaya in 1938. Currently, STFA is one of Turkey's most well-established and reputable companies providing services in the construction and construction equipment sectors.
   www.stfa.com

TBMM Library: TBMM Kütüphanesi
   The Library of the Turkish Republican Parliament.
   https://acikerisim.tbmm.gov.tr/

KGM publications: Karayolları Genel Müdürlüğü
   KGM is the current authority for the road transportation network of Turkey. It was founded formally in 1950 as a continuation of Nafia.

İU Library: İstanbul Üniversitesi Kütüphanesi
   İstanbul University Library – Rare Works Collection

Chapter 1

Introduction

1.1 INTRODUCTION

Turkey's current borders were established in a peace treaty negotiated during the Lausanne Conference, signed on 24 July 1923, which followed the victory of the independence struggle led by the nation's leader Atatürk.[1] This pared-down boundary was the outline of a new country, Turkey, founded as a republic after many centuries of the sultanate of the Ottoman Empire. The new republic flourished with many revolutions in legal, educational, social, religious and economic areas, like the change of alphabet from Arabic to Latin letters, and the adoption of the secular regime that abolished the caliphate that had been considered the sultan as a religious head. The Republic of Turkey, referred to as the "Rebirth of a Nation",[2] was in fact a complete transformation for its people from being the Sultan's subjects to then becoming citizens of a modern republic with civil rights and political equality for all including minorities and women.

This new era witnessed the start of many technical, cultural and social developments as the new regime progressively worked to build its nation. The empire had fallen behind contemporary development, especially "missing the train" of the Industrial Revolution, leaving the country mostly undeveloped and in a rural state after more than 10 years of war and also with a considerable debt to foreign countries. At the same time, the young and educated generation was cultivated, and the established government system benefited from the latest reforms of the empire. Therefore, the new republic can be described as "a rebirth of a nation from its ashes".

This book will introduce the bridge-building in this new era including technical and heritage features within the context of engineering history, and with the soul of the era as its background. I call these bridges "Köpriyet", a name derived from Köprü (bridge) and Cumhuriyet (Republic). Even though the first years of the Republican era can easily be isolated with its own rich and unique content, this book also introduces the history of bridge-building in the region, thus providing a better understanding and connection to the aforementioned period.

This book is written with "a wish to compose a history of these bridges, cherish their value and memories"[3] as directly quoted and translated from Çulpan's preface in his book "Türk Stone Bridges".

The first chapter of the book starts with a brief history, and then, the content of the chapters is described. The resources and the methodology of the study are explained, and the information regarding the main resources is referenced.

DOI: 10.1201/9781003175278-1

The second chapter provides an introduction to the country and to bridge-building history in Turkey. In this chapter, the emphasis will be given to the late Ottoman period and the conditions established with the reforms during the empire. The transportation infrastructure works gradually transformed their focus from mainly military towards strategically built bridges for public service.

Chapter 3 explains Republican engineering and bridge-building conditions in terms of technical, organizational, methodological and social policies. The engineers who worked in bridge-building and transportation infrastructure projects in the new country are briefly mentioned. Their story alone is a social phenomenon that deserves another book.

Railway engineering and building of rail infrastructure are also briefly mentioned in Chapters 2 and 3. Railway engineering had a leading role and was parallel with the building of road infrastructure during the Turkish industrial transformation; therefore, it was included to complete the story. Even though railway bridges are not covered in the content of this book, they are not ignored and some of them are mentioned where information is available for these bridges.

The next Chapters 4–7 will introduce bridges which are grouped according to their structure types, building materials and significant features. These chapters namely devoted to Bowstring bridges, Concrete Arch bridges, Metal bridges, and Beam and Unique bridges.

Chapter 4 mentions the international history of this type as bowstring concrete bridges are rare anywhere outside the USA. Chapter 5 focuses only on the early Republican era to narrow the content. Chapter 6 will also mention the first applications of their type in the Ottoman period. Finally, Chapter 7 discusses beam bridges which are significant or representative of their type. Unique bridges, also introduced in Chapter 7, contain bridge with the sole or very rare applications of their types.

The material and type-specific histories of the bridges are included in each relevant chapter.

Finally, Chapter 8 will provide a summary for the bridge-building of the early Republican era.

1.2   RESOURCES AND CONTENT OF THE BOOK

The resources of this study were collated from various documents and reports. The new republic was comparatively productive in terms of providing imprints as the new nation required to be recognized nationally and internationally. For example, the tenth year of the republic had been celebrated with formal ceremonies and many publications were created to show its progress and achievements.

Therefore, a report[4] was published for the tenth anniversary of the Public Works Authority/Nafia.[5] This organization was established to construct all public structures, including irrigation, ports, railways and roads. The report provides detailed descriptions for the bridges built between 1923 and 1933. The bridges constructed by public works including the ones constructed and handed over to local provinces (Vilayet) are described as totalling 48 bridges in this report.

Another source used was the list provided by Örmecioğlu.[6] This list, covering bridges from 1924 to 1960 and collated from numerous resources, totals 231 bridges.

This source was intended to list all the bridges built by the General Directorate of Highways (KGM).[7]

Neither of the resources specifically included the bridges built by provinces and municipalities. The bridges built by local governments were found in a separate report named "Provincial Works Album[8]" which covered a range of infrastructure works including schools, hospitals and bridges. The bridges in this report are mainly timber, stone and short span beam bridges. Municipality bridges, on the other hand, are not covered by any publications; such infrastructure works are included in this book where resources were available.

KGM was officially founded in 1950, and after this date, there was a construction boom in road- and bridge-building. Two main KGM publications are used in this study as a source to identify the bridges. The first one is "*Köprülerimiz* (Bridges)",[9] which lists and provides photographs of some of the bridges from 1950 to 1968 where the total length of the structure is more than 80 m, and the second publication is titled "*Yol Ağlarımızdaki Köprülerimiz* (Bridges on our Road Network)",[10] which lists all the bridges and photographs of some of the bridges over 100 m.

Combining available lists with field surveys of the author, 24 bowstring bridges, 54 arch bridges, 14 metal bridges, and 20 beam and unique bridges were identified and included in this book. The total number of the beam bridges was around 150; however, only some with some special characteristics or representing typical examples have been presented. The stone and timber bridges have generally been excluded to limit the scope of the book.

Essential information about the bridges is found in the periodic publications made by Nafia. These periodic magazines were published to cover large-scale public infrastructure works such as bridges. A thesis study of Haykır[11] combines the information regarding transportation infrastructure in these magazines.

Another important complementary source for this study was the memoirs of eminent engineers written as biographies of their works. One of the earliest is Atayman,[12] an Ottoman engineer, who worked on the Hedjaz railway project and later in railway projects in the early Republican years. The other main document providing a live insight into Köpriyet was by Feyzi Akkaya who narrated his works in a book called "Kilometer Stones of Our Life[13]".

As the content of the book expanded and therefore had to be controlled, a sad exemption had to be made for the engineers themselves, who built these bridges and exemplified the soul of the era. The original plan included a dedicated chapter on the lives and achievements of these pioneers, especially focusing on the conditions in which they worked. Nonetheless, Kemal Hayırlıoğlu, Halit Köprücü, Feyzi Akkaya and Sezai Türkeş, four prominent engineers of the early Republic, are introduced in Chapter 3. They were part of the Köpriyet since the beginning of the era and had been involved in building many of the significant bridges of the Republic. For the other engineers, who were involved in bridge-building, their short biographies have been given throughout this book, mostly in endnotes, whenever the information was available to the author.

The end of the Köpriyet era is not marked with any specific date or event; however, the year 1933 is selected since the complete bridge lists were only found until this date. Beyond this year, only some significant bridges with social and technical significance are included.

Inventory Study: This study included a considerable amount of work for gathering an inventory of the bridges. The main reference for the inventory was Nafia reports and the list provided by Örmecioğlu. The lists were collated and modified with revisions and contributions from other resources. The listing in this book includes all the bridges found from available sources; exemptions are the railway bridges, stone and timber bridges.

The inventory will be given in the relevant chapters of the book. This inventory aims to include as much information as possible within the limited space, and the coordinates of the bridges are also added to assist in finding them on maps.

The list of works made in local provinces and municipalities, by the nature of the information available, are not considered entirely complete due to the availability and significance of such structures. In addition, as the Ottoman technical archives for foreign works were recorded mainly in French and the first years of the Republic were made in Arabic, such documents that remain untranslated have not been used here.

Therefore, the scope for further study remains to include these and other sources in future.

Photographs: One of the challenges of writing such a book was to source images of bridges, as photographs and sketches are vital to describe the bridge. In this way, it was easier to write in a blog; however, a book is constrained by copyright requirements where the original source included such images.

The majority of the bridges are also described in the above-mentioned Turkish language blog: www.köpriyet.blogspot. The photographs of bridges can be seen at blog posts.

When Reading the Book: In this book, the bridges are identified with the codes such as Bowstring – BWS, Concrete Arch – ACH, Metal – MTL, and Beam and Unique Bridges – BRD, followed by a number which is its location in the chapter. Therefore, whenever a bridge is mentioned throughout the book, its coding is also given so that the reader can find the bridge without a need for further guidance. When a bridge is mentioned within its own chapter, this code is not used in order to simplify the text.

Terminology: Many of the main resources of the book were dated the 1930s; therefore, some of the terminology used back then is outdated. Therefore, instead of direct translations, the up-to-date terms are used in English text. For example, the use of the term "demir" to refer to both iron and steel was common in original descriptions, and later, the terms "çelik" and "demir" were adopted for steel and iron, respectively. Where the type of the material is known, then the material is specified; otherwise, the term metal is used to refer to both of the materials.

Exceptions of the Study: As mentioned in related chapters of the book, masonry bridges and timber bridges were generally exempted from the content of the book.

Another exemption was the financial aspects of the bridge designs. However, the costs of nearly all structures were recorded often in detail in reference documents. This has generally been omitted from the book in order to limit the content.

NOTES

1. The boundary was finalized except for the province of the Hatay in south-eastern Anatolia, which was acquired after a referendum in 1939.

2. Kinross, L. (1964) *Atatürk: The Rebirth of a Nation*. Weidenfeld and Nicolson. London.
3. "köprülerin bir tarihçesini meydana getirmek, değer ve hatıralarını yasatmak dileğidir"
   Çulpan, C. (1975) *Türk Taş Köprüleri*. Türk Tarih Kurumu. Ankara.
4. Vekaleti Nafia, T. C. (1933) *On Senede Türkiye Nafiası 1923-1933*. Nafia Vekaleti Neşriyatı. İstanbul.
5. The term used for public works was "imar", which means civilization and development excluding the social content. The term later changed to "Nafia", which can be translated as utility or beneficial actions, to refer to public assembly at the governance level. Nafia also later changed to "Bayındırlık", which means prosperous, in 1935.
   Tekeli, İ., and İlkin S. (2004) *Cumhuriyetin Harcı: Modernitenin Altyapısı Oluşurken*, Bilgi Üniversitesi. İstanbul. Page: 5.
6. Örmecioğlu, H. T. (2010) Technology, Engineering, and Modernity in Turkey: The Case of Road Bridges between 1850 and 1960. Doctorate thesis submitted to Middle East Technical University (METU). Appendix C: The list is collated from the State National Achieves, Ministry of Public Works, KGM archives, Bayındır Magazine and KGM bulletins.
7. General Directorate of Highways (Karayolları Genel Müdürlüğü, KGM) current authority responsible for highways in Turkey.
8. Province Album (1929) *Vilayet-i Hususi İdareleri Faaliyetlerinden*. Hilal ve Cumhuriyet Matbaaları. İstanbul Metropolitan Municipality Taksim Atatürk Library (IBB Atatürk Library) Album name 454 big and 2238 small bridges and roughly 120 bridge/culverts are photographed.
9. KGM (1969) *Köprülerimiz*. KGM Matbaası. Ankara.
10. KGM (1988) *Yol Ağlarımızdaki Köprülerimiz*. KGM Matbaası. Ankara.
11. Haykır, Y. (2011) Atatürk Dönemi Kara ve Demiryolu Çalışmaları. Ph.D. thesis submitted to Fırat University.
12. Atayman, M. S. (1967) *Bir İnşaat Mühendisinin Anıları, 1897-1918*. Baha Matbaası. İstanbul.
13. Akkaya, F. (1989) *Ömrümüzün Kilometre Taşları: STFA'nın Hikayesi*. Bilimsel ve Teknik Yayınları Çeviri Vakfı. İstanbul.

Chapter 2

Turkey History

2.1 BRIEF INTRODUCTION TO TURKEY

The territory of Turkey is roughly a rectangular shape of approximately $1600 \times 800\,km$ dimensions in east-west and north-south directions. The land is bounded on the north by the Black Sea, on the west by the Aegean Sea and on the south by the Mediterranean Sea. It shares land borders with eight countries.

Turkey has 93% of its landmass in Asia and the remainder in Europe. It is geologically very diverse with various climates, different flora with drylands on the eastern side and fertile land on the western side, many rivers of different sizes, and rich geology like gorges, large meadows and volcanic regions. Anatolia is a name commonly given to the area of Turkey that is part of the Asian continent. This region is also called "Asia Minor".

Turkey has several long rivers. The Fırat (Euphrates) and Dicle (Tigris) rivers, both originate from Anatolia, flowing from north to south within Turkey for 1250 and 530 km, respectively, before entering Syria and then Iraq, are the world's 20th and 65th longest rivers. The rivers Fırat and Dicle outline Mesopotamia, the "Land Between the Rivers". The longest river in Turkey is Kızılırmak (Halys) with 1355 km, all within the country's boundary. However, Turkey's rivers are not very wide compared to the world's other rivers (Figure 2.1).

Anatolia has been home to the oldest permanent settlements in the world, like Göbekli Tepe, the oldest known man-made structure.[1] Since then, many civilizations have settled in these lands including the Greeks, Helens, Romans, Seleucids, Anatolian Beyliqs and lastly Ottoman Empire before the foundation of the Turkish Republic. These civilizations introduced different practices, traditions and cultures, which accumulated throughout the centuries.

Moreover, Turkey has a strategic location as a transcontinental country with two narrow straits that connect Asia to Europe. Historically, these straits, the Bosporus and the Hellespont (Dardanelles), had been first crossed with floating bridges made with boats connected with ropes and flax by Persians Kings Darius in 513 BCE and his son Xerxes in 480 BCE, respectively.[2]

The south-eastern region was part of the "Fertile Crescent", where settled farming first emerged as people started the process of clearance and modification of natural vegetation to grow newly domesticated plants as crops. This led to technological advances including the development of irrigation, the invention of writing, the wheel and glassmaking.[3]

DOI: 10.1201/9781003175278-2

Figure 2.1 Map of Turkey showing cities and rivers.

The geographical location of Turkey is often described as a "bridge" from Europe to Asia or from Occident to Orient. Ancient roads have existed in Turkey with the oldest known as the King's Road which carried a postal service from İzmir (Smyrna) to Persia. The military routes to west and east, the trade routes such as the Silk Road and Spice Road to Asia and the Pilgrimage route to Mecca all remained important until the advances of the 19th century diverted major transportation to sea routes and later railways.

2.2 ANATOLIA BRIDGE HISTORY

The earliest known bridge[4] was built by the Hittites, as a timber log structure bedded in the natural rock and crossing over a deep gorge. Another early crossing was an ancient stone clapper bridge in Assos with special engineering features such as wooden clamps and dowels as well as notches to prevent slippage of stones.[5]

An interesting example is the 850 BCE dated bridge in İzmir registered in the Guinness World Records (2017)[6] as "the oldest datable bridge in the world still in use" even though it is hard to substantiate other than all the bridge history books agreeing on it. It is also called Caravan Bridge, having carried the caravans over the River Meles, known as the birthplace of Homer. The current bridge is a single 8.5 m span made of stone.[7]

More than 70 bridges dating from Roman to Hellenistic times are scattered around Turkey along the Roman roads.[8] Many have distinctive features such as Çavdarhisar (Aezonoi) and Bergama (Pergamon) bridges still surviving with their original architectural details. For example, the Karamağara Bridge has each of its voussoirs carved with letters in Greek that say: "...Lord God may guard your entrance and your exit from now and unto all time..."[9] Dimçayı Bridge has three rings for its arch, Limyra

Figure 2.2 Çavdarhisar (Aezonoi) Roman Bridge (author).

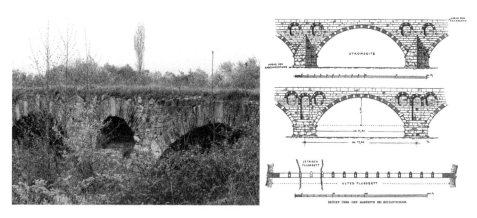

Figure 2.3 Left: İnikli Roman Bridge (author). Right: Sultançayır (Makestos) Bridge (Wiegand, 1904).

Bridge is a shallow segmental arch, and İnikli Bridge has alternating brick and stone voussoirs for its arch (Figures 2.2 and 2.3).

200 BCE dated Cendere Bridge has an impressive span of 34 m with 7.5 m width but interestingly only 4.5 m clear passage between the parapets. The parapets are configured to keep a pattern and to comply with the gradient of the bridge. They are stopped towards the ends with pedestals, one of which is used for inscriptions (Figure 2.4).

Cendere Bridge has a distinctive feature of columns of 9–10 m height, one each at either side of its entrances. Each column was dedicated to a family member of Emperor Septimius Severus and held a statue of that person. The ones on the south side are dedicated to Septimius and his wife Julia Domna, and the other two on the northern side to their sons, Caracalla and Geta. Cendere Bridge tells us a classic story of the battle

Figure 2.4 **Cendere Roman Bridge (Efkan Sinan/Flickr).**

Figure 2.5 **Cendere Bridge from the 1970s (SALT research, M. Erem Çalıkoğlu archive).**

for the crown. The two sons of the Emperor, Geta and Caracalla, were always in strong opposition to each other. Caracalla ended these arguments by murdering his brother. Consequently, Geta's column was removed from the bridge and his name was chiselled from the inscription to delete his memory from history (Figure 2.5).[10]

Among many aqueduct bridges, Polio Aqueduct Bridge is an outstanding example dated through its inscription at 4–14 CE (Figures 2.6 and 2.7).[11]

Later occupation of Anatolia by Persians introduced different styles and technology. Malabadi Bridge from the 11th century has an incredible span of 38.6 m and allowed for ships to travel the Fırat (Euphrates) River. Another interesting feature of the Malabadi Bridge is the rooms built in either side of the spandrel fill; the rooms are accessed by stairs, and facilities were provided in them for travellers (Figure 2.8).

Figure 2.6 **Polio Aqueduct Bridge dated 4-14 CE (Choiseul-Gouffier, A., 1823).**

Figure 2.7 **Polio Aqueduct Bridge (Efkan Sinan/Flickr).**

There have been many civilizations such as Seleucids and Beyliqs before the Ottoman Empire, all built bridges; however, the difference is not always easily recognizable between them as the building technology relied on local materials and local workforce, experience and construction methods. For example, the Seljuk bridges were often irregular and eclectic, as they were regularly constructed over the remains of previous bridges. Their bridges do not necessarily have any order, symmetry or straight lines. The only certain identification has been made where inscriptions are available (Figure 2.9).

Anatolia had local construction techniques developed significantly over the ages with the abundance of timber and stone. Timber was common, but the remaining

Figure 2.8 Malabadi Bridge (author).

Figure 2.9 Adala Bridge over Gediz River in Manisa (author).

examples are extremely rare. Some outstanding timber bridges chosen below are, for example, Yenice covered bridge with timber truss,[12] and the other is a King Post Truss resting on cantilevered logs extending from stone abutments. Both bridges are constructed by local people and repaired many times after floods (Figures 2.10 and 2.11).

A rather different collection of bridges are found in the Black Sea Region where the hills rise quickly from the coast and settlements are divided by frequent rivers. Bridges are built locally in this region, and some have daring spans. A few of these bridges have chains hanging below them from their keystones. There is not yet a convincing explanation for this chain. It was suggested that they were carrying a stone with the bridge date

Figure 2.10 Yenice Timber-covered Bridge dated 1932 (author).

Figure 2.11 Deretam Timber Bridge with King Post Truss Span (author).

Figure 2.12 **Stone Bridge from Black Sea Region of Turkey with hanging chain below the crown (author).**

written on it, or that it was utilized to apply weight on keystone to place it at the crown. This stone might also have been working as damping in modern engineering terms, reducing the displacement of the bridge by the weight provided by the stone (Figures 2.12 and 2.13).

2.3   OTTOMAN BRIDGE-BUILDING

Romans marked their roads with milestones in Anatolia. Then, Seljuks and later Mongols had a road system which was secured by traveller's lodges called "ribat" and later "caravanserai/han" at certain intervals. Subsequently, these secure roads facilitated trade while taxation of the road users provided resources and impetus to improve the network. The Ottomans inherited an extensive road network and organization from its predecessors.

The Ottomans improved this system and built new bridges and caravanserai. Road- and bridge-building was already a profession in these times. The maintenance of these bridges and the road network was provided by a well-established Ottoman system known as the "derbend organization".[13] In this organization, "derbend" was founded in strategic and dangerous crossings, usually a bridge. The officers for road-building and maintenance were called "kaldırımcı", and bridge builders and maintainers were

Figure 2.13 Detail of hanging chain below the crown (author).

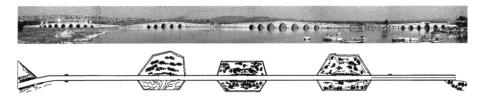

Figure 2.14 Büyükçekmece Bridge 4 bridges joined with islands (author).

called "köprücü/meremmetci", and both had to be registered. They were usually chosen from among the local villagers or townspeople in the vicinity.

The Ottomans built some remarkable bridges like the world's longest stone bridge with 174 spans in Uzunköprü.[14] They achieved marvels like the Büyükçekmece Bridge and the Mağlova Aqueduct Bridge both designed and constructed by Sinan and representing the period of the architectural peak of the Ottoman reign. Büyükçekmece Bridge has four separate bridges built on three man-made islands. The first bridge has balconies, and the last bridge has inscription towers. The bridge entrance has bollards on both sides, and even birdhouses are embedded in spandrel walls. A characteristic representative of the Ottoman bridge-building can be considered as the 1663 dated Babaeski Bridge, with its architectural and technical features (Figure 2.14).

Figure 2.15 Mağlova Aqueduct Bridge (author).

Ottoman bridge-building was done mainly in stone often with metal clamps and ties known to be used for structural connections. The foundations are generally built as a grillage system which consisted of timbers laid out perpendicular to each other in two layers which are then filled with gravel (Figure 2.15).

The decline of the Ottoman Empire can be highlighted with a bridge over the Meriç (Maritsa), in Edirne, since it was the last large-scale bridge and its construction took over 10 years probably because of the struggle to finance it. The 1847 dated bridge shows the features of Ottoman bridges with cutwaters, bollards, drainage details and especially an impressive artwork that is the marble inscription tower and balcony at its mid-span (Figures 2.16 and 2.17).[15]

2.4 OTTOMAN PUBLIC WORKS PERIOD

The Ottomans had a Corps of Royal Architects (Hassa Mimarlar Ocağı), which was established in the 15th century, as a military organization responsible for construction activities such as design and management for all structures including bridges and roads. The corps was also responsible for the repair and maintenance of structures that they built and to assure road access before military campaigns including bridges. They managed the work outside the capitol through state architects. They were mainly engaged in military service construction with the actual bridge-building carried out by local organizations, called "eyalet/province" and "sancak/liva".

The finance of the construction works was carried out by Şehremini (City Prefect), whereas the salary and number of workers were determined by Corps Architects. Later, these two establishments merged during the reforms of Mahmud II in

Figure 2.16 **Meriç Bridge in Edirne (author).**

Figure 2.17 **Meriç Bridge inscription tower (author).**

1831 under the name of Imperial Building (Ebniye-i Hassa).[16] Then, in 1838, this was replaced by a Council of Beneficial Works (Meclis-i Umur-ı Nafıa) established to cover all public works.

Many reforms and reorganizations were initiated in an attempt to keep pace with the industrialized world, especially starting with the 1839 Imperial Edict and constitutions. However, the Empire could not stand long against the progress of the industrial revolution in technical matters and the influence of the French Revolution on social issues.

Events, like wars, reforms and establishing constitutions in parallel with educational changes had an impact on government organizations, public works activities especially acting a driving role in development and industrialization. Therefore, the events together with changes in engineering education,[17] public works establishments and programmes, and the new codes and regulations are given in Table 2.1. Public taxation was also added to the table as it was the main financial resource in carrying out the activities. Then, the last column shows some significant public works activities.

The reforms and reorganizations were either supported or pushed by foreign investors, mainly Britain, France and Germany. These imperialist powers of the industrial revolution needed raw materials easily transported to their factories for mass productions and also a market for their products. Consequently, these investors built roads and railways and opened small- to large-scale factories in Anatolia. Road projects were smaller and required less technology; thus, they were often carried out by Public Works Authority/Nafia[18] except for the engineering services required.

Most of the engineering education reforms and activities were carried out by foreign engineers from Britain, France and Austria. For example, French engineers[19] were first involved in some fortification projects and later in military projects and then education. For the period between 1854 and 1865, at least 47 foreign engineers were identified as appointed by the Ottoman authorities to work on different civil projects, in state institutions, or for provincial and local administrations.[20]

In terms of public works and activities, in 1845, Nafia Treasury was formed and parliament established Civil Councils (İmar Meclisleri) which organized local districts and had delegates in parliament. These councils then prepared an inventory of required investments for the public works. In 1848, public works were organized separately as a Ministry under the name of Ministry of Public Works (Nafia Nezareti) to form a central authority for urban planning and regulations. Nafia was responsible for all kinds of public utility works in terms of transportation infrastructure, road and bridge constructions, city roads and railways.[21]

For the cities, road-building and its management were carried out by municipality (İhtisab/Şehremaneti), and if the scale of the work and also the urgency required it, then Nafia became involved in these works as well.

In parallel to the activities in the field, regulations were put in place to actualize the projects and works, also providing the necessary set-up for further improvements. In 1839, "Official Record" was the first urban regulation formulated for the capitol İstanbul. This regulation focused on the reconfiguration and widening of the street network as an inevitable result of the big fires that devastated the mostly timber-built city. The primary roads, such as Divanyolu, the protocol road from Topkapı Palace to the city wall on the west, and the coastal roads on both sides of the Golden Horn would

Table 2.1 The Events, Educational Reforms, Nafia Establishments, Regulations and Public works Activities (author)

Events	Engineering Education Reforms	Nafia Establishment and Programmes	Codes and Regulations and Taxation	Public Works Activities
1839: Imperial Edict of Reorganization	1839: First civil university for engineers opened but did not survive long	1838: Council of Public Works (Meclis-i Umuru Nafia) was established and ceased later. 1845: Formed again	1839: İstanbul City Road Plan Official Record	1834: The vehicle road Üsküdar-İzmit Postal Road constructed 1839: First road design in İstanbul by Moltke 1850: First Nafia Roads started Bursa-Mudanya, Bursa-Gemlik, Trabzon-Erzurum 1856: İzmir-Aydın Railway started
	1848: Land Engineering School was divided into two sections: one for fortification officers/architects and the other for artillerymen	1848: Public Works Ministry (Nafia Nezareti) established	1848: Building Regulation	
1853-1856: Crimean War 1856: Imperial Reform Edict			1856: Roads and Bridges Regulation 1863: Street and Building Code 1866: New Roads and Bridges Code – Road tax defined 1870: Road Taxation System revised 1898: New Roads and Bridges Code amended	
1875: Declaration of Bankruptcy 1876: First Constitutional Period 1878: Suspension of Cons. By Empire	1874-1879: The Imperial Lyceum of Galatasaray (Mekteb-i Sultani) offered Higher education in civil engineering adopted the model of "Ponts et Chaussées" 1883: Engineering School (Hendese-i Mülkiye Mektebi) was a quasi-military institution. This school adopted the model of "Ponts et Chaussées" and later transformed it to a German model.	1882: First Nafia Plan by Hasan Fehmi Pasha		1900-1908: Hedjaz Railway of 1200 km

(Continued)

Table 2.1 (Continued) The Events, Educational Reforms, Nafia Establishments, Regulations and Public works Activities (author)

Events	Engineering Education Reforms	Nafia Establishment and Programmes	Codes and Regulations and Taxation	Public Works Activities
1908: Young Turks Revolution and Second Constitutional Period	1909: Engineering School was removed from military supervision and established as the Higher School of Engineers (Mühendis Mekteb-i Alisi) under the Nafia.	1908: Second Nafia Plan by Gabriel Noradounghian		1909: Regie was given a concession for all roads construction and maintenance
1914-1918: World War I	1912: Robert College Engineering School established the civil engineering department. Later transformed to Bosphorus University.		1914: Road taxation revised	
1915: Gallipoli War				
1918: Armistice of Mudros and Occupation of İstanbul				
1920: Treaty of Sevres and occupation of Anatolia				
1919-1922: Turkish War of Independence	1922: Nafia Fen Mektebi was founded for educating technicians adopting the model of Ecole de Conducteurs. Later transformed to Yildiz Technical University.	1923: Third Nafia Plan by Fevzi Bey	1921: Road taxation revised	
1922: Armistice of Mudanya	1928: Civil Engineering School has been transformed to Higher Engineering School (Yüksek Mühendis Mektebi). Later transformed to İstanbul Technical University	1929: Fourth Nafia Plan by Recep Peker		
1923: Treaty of Lausanne				
1939-1945: World War II			1948: Cease of road taxation	

be 15 m wide with 3 m wide pavements on both sides. In addition to the primary roads, the roads were further divided into categories with a defined width.

In 1866, the Road and Bridges Regulation classified the road system into categories according to their strategical use and defined the width, slope, construction method and typical sections.

The transformation was also observed in terminology; for example, "kaldırım"[22] was previously used for the stone-paved portion of the whole road section, and later, when the road was reconfigured to be shared by vehicle and pedestrians, "kaldırım" referred to the sidewalk on either or both sides of the road. Then, the name "şose" imported from the French "chaussée" was used to define the vehicle road which had a gravel finish.

In 1875, the Empire declared bankruptcy, and especially after this date, codes, regulations and activities ceased except the projects that were carried out through concessions. The engineering projects were mainly directed by foreign investors. During this period, local authorities constructed some roads with their limited resources. Some local governors achieved considerable results, while the capitol İstanbul had many engineering projects through concessions. Meanwhile, road-building consisted mainly of maintenance and repairs of the transportation system with mostly timber and some stone bridges.

In 1882, the first Nafia programme was prepared by the Ministry of Public Works and was submitted as a detailed report to the Sultan. The Minister's report envisaged 2535 km of road, and one of the financial models for road construction was the foreign capital in the exchange for concessions.[23] The Minister, Hasan Fehmi Pasha, stated it was necessary to attract foreign engineers and capitalists to exploit the huge assets of agriculture, forests and mineral resources of the Empire.[24]

In 1908, the second Nafia programme was prepared by the Minister of Commerce and Public Works, Gabriel Noradounghian, and presented to the grand vizier with a balance of existing infrastructures in the Ottoman Empire and a Public Works programme to improve and expand them. As analysed in detail by Tekeli and İlkin,[25] this programme was based on previous research, part of it done by Ottoman engineers and officials of the Public Works Administration, and included comparative data situating the Ottoman Empire in the context of other countries. The programme covered roads, railways, ports and hydraulic works (canals, irrigation and draining of marshlands).[26]

The end of the Empire came with the Young Turks Revolution, which brought the necessary mental change for the transformation from the "subjects" of the Sultan to a society of "citizens" of the Republic.

The Committee of Union and Progress (CUP),[27] a revolutionary organization founded in İstanbul and later aligned with the Young Turks,[28] forced Sultan Abdülhamit to reinstate the suspended constitution in the Young Turk Revolution. CUP itself, while it did not actually form a party, issued a general manifesto of its policies and supported those candidates who promised to follow it, thus forming them into a group that came to be known as the Unionists (İttihatçılar).[29] The railway projects continued, especially Hedjaz and Baghdad railways, and the sewerage system of İstanbul was completed. CUP did not seem to achieve many changes; however, it brought about a generational change in the system and replaced many officials with young ones and also aimed to create a local bourgeois and elite.

Road construction, with postal, trade, other transportation and military services, had always remained a priority, whereas the scarcity of resources, budget and expertise together with the unstable conditions within the empire and organization did not lead to any improvements in road-building despite reforms up until the decline of the empire. On the other hand, the local organizations stayed outside this environment and road-building continued in these locations with existing and entrenched methods. Therefore, the bridges constructed during the decline of the empire were mainly medium-scale stone and timber bridges built locally. Some significant examples, although still only of medium scale, can be found which were carried out mostly by the foreign engineers employed in the organization.

In order to provide a closer look at the period, some notable projects related to bridge-building, mainly railway projects and road projects, will be described in the following pages.

Railway Projects: The railway itself could be described as the pre-eminent subject of modern technological advancement of the time. It was an essential instrument for trade required by foreign European colonial countries operating within Anatolia. Indeed, the railway projects themselves were good investments with guarantees and concessions provided. The long-term risk was eliminated from railway projects with the so-called kilometre guarantees, which was compensation guaranteed by the Empire in case railways did not provide profit to investors for a certain period. Short-term profit of the railway projects was ensured by the right to own and use all the resources of mining including oil, forest, even archaeological finds and extending to the production in agriculture within a certain corridor on either side of the lines. In addition to all, state-owned lands, buildings, warehouses, etc., were allocated to the use of foreign investors and they were also free of import duties.

For the empire, the railway was a beneficial tool to strengthen its control and power and to speed military access to regions where especially heavy taxes were fermenting revolts.

The earliest known railway proposals started in the 1830s when the British were trying to find transportation routes to the east. Chesney[30] carried out an investigation of the Fırat (Euphrates) River for its steamboat use. According to 1857 dated Chesney's report,[31] river transport could be connected to the Mediterranean Sea by railway; however, the proposal did not proceed.

The first railway construction started in the Empire in 1851 with the Alexandria (İskenderiye)–Cairo (Kahire) line, which was built by the British and completed in 1856.[32]

The first railway lines that stayed completely within the current borders of Turkey were the İzmir–Aydın and İzmir–Kasaba (Turgutlu) lines. They were built between 1856 and 1912 by British and then French companies, and both lines had a 6% profit guarantee for 50 years. İzmir–Aydın line also had a 45 km corridor for the resources concession. These lines further extended to Bandırma, Afyon and Eğirdir. The layout of these lines was configured as branches of the tree meeting its root at the harbour for transporting goods for import and export of the goods and material (Figure 2.18).

Another line was the Rumelia Railway line constructed between 1869 and 1888, later called Şark (Oriental) railways, with 468 km[33] within current Republic borders. Line

Figure 2.18 1896 dated railway bridge on Alaşehir-Uşak extension of İzmir-Afyon Line (İBB Atatürk Library).

concession was given to Baron Hirsh, who constructed the line with the cheapest possible methods. Bridges solely were made of timber, rather than metal superstructure and stone substructure, and the sleepers were made of pine timber instead of oak.[34] The Sultan had started an investigation for the project during the handover after its temporary acceptance. Thomas Page[35] was involved in this investigation.

Rumelia lines were a dreadful experience for the Empire, because of its high cost, long duration and the poor outcome combined with the disputes with Hirsch and the corruption within the empire authorities. Therefore, the Empire decided to build lines themselves. The 42 km Mudanya–Bursa line was started in 1873 and finished in 1875; however, because of the poor quality of the construction and insufficient funds to operate the line, it was leased to a British Company.[36] The Haydarpaşa–İzmit line with 91 km started in 1871 and was completed in 1873; however, the Empire treasury was bankrupt and the line operations were ceased.

Wilhelm Von Pressel[37] was assigned as a chief engineer of the Empire in 1872 to prepare the projects for railway lines (Figure 2.19).

Among many other rail lines that mainly carried resources to ports, Baghdad and Anatolian rail lines were different being strategic and continuous throughout the country. These lines were extending from capitol İstanbul to Eskisehir and then were continuing in two branches as one to Ankara and Kayseri and the other to Konya, respectively. Later, both lines met at Diyarbakır and continued to Baghdad.

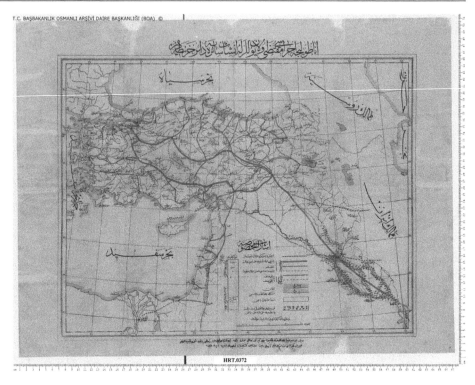

T.C. BAŞBAKANLIK OSMANLI ARŞİVİ DAİRE BAŞKANLIĞI (BOA) ©

HRT.0372

Figure 2.19 1874–1883 dated map of proposed rail lines and roads within Anatolia (BOA HRT.h. 372).

An 1896 dated map[38] prepared by Pressel shows the Baghdad line extending as a separate line from İstanbul to Baghdad. However, the Anatolian line had to cease at Ankara.[39] The Baghdad line was later shifted to Konya and Adana before continuing to Baghdad. Both lines were constructed by Germany who dominated the railway projects in the Empire for years after. Strangely, the operational language was still in French even on these two rail lines (Figure 2.20).

The Anatolian line had many long bridges like Vezirhan, Karasu, Pekdemir, Başköy and Beylikköprü between Adapazarı and Ankara crossing over Sakarya River and its branches. Manas Efendi,[40] an engineer working in Nafia, wrote a very useful magazine article[41] introducing the ancient and modern bridges crossing over the Sakarya River.

These lines were damaged significantly during the Independence War. The repair works of the Beylikköprü line were carried out by technical staff brought from the Belemedik section, which was a challenging crossing of the Baghdad line near Adana. The Başköy viaduct was temporarily built as a timber trestle bridge using the sleeper timber of the rail line.

The Baghdad line, also known as Orient Express, faced challenges such as crossing over mountains and the Fırat (Euphrates) River at Jarabulus. Jarabulus Bridge (also called Cerablus and Karkamış) is at the border with Syria. The Djihan (Ceyhan) and Seyhan bridges, which cross rivers of the same names, still survive within the current

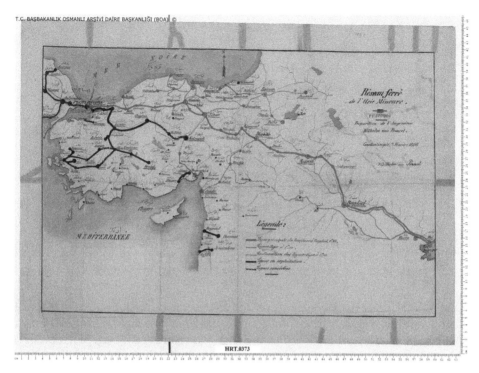

Figure 2.20 1896 dated map to show proposed lines by Wilhelm von Pressel, legend from
top to bottom: Bosporus–Baghdad Mainline at 1.43 m, narrow-gauge lines at
1 m, continuation of the narrow-gauge lines at 1m, lines in operation and lines
approved (BOA HRT.h. 373).

borders of Turkey. Hereke Trestle Bridge with 85 m height is in Syria nowadays, 1 km
away from the Turkey border. Another bridge is the famous Hacıkırı Bridge in Adana,
also called Gavurdere Bridge, an ashlar masonry viaduct with 69 m height. These
bridges are briefly described in an article by Gordon Heslop (Figure 2.21).[42]

During the Independence War, before 23 April 1920,[43] the Lefke, Geyve and
Cerablus bridges were demolished by the orders of Atatürk. These bridges were
important control points on the western and eastern frontiers (Figure 2.22).

An influential project was the Hedjaz Railway (1900–1908), which is sometimes
referred to as the initiator of World War I, started by Abdülhamit II for the connection of İstanbul to the "Sacred Lands" in the Arabian Peninsula. The project
caused a political power struggle between investors. The Ottomans planned to
construct the line with local resources, however soon realized that the experience
of the local engineers was not yet sufficient for the project. The project was handed
over to Germany, and Meissner Pasha[44] became the head of the project. Muhtar
Çilli[45] played an important role in the project, especially for the section in the Holy
land where only Muslims were allowed to enter. This project provided essential
experience for local engineers.

Figure 2.21 Hacıkırı (Gavurdere) Bridge on bagdad railway line (author).

At that time, it was compulsory for engineering school graduates to work on this railway project. The challenges of working in desert conditions were primarily water access and storage, dealing with sand storms and accumulation of sand along the rail line. There were many bridges on the line, and a stone bridge with two tiers at Amman is one of the fascinating examples for bridges. The resources in the vicinity were only stone, and the bridges of this line were mainly stone and metal bridges. The bridges still had to be designed for flood even in the desert as the floods could be devastating, mainly because of the poor soil infiltration.

Atayman,[46] an Ottoman engineer, worked on the Hedjaz Railway. He was involved in some bridge repairs on this line during World War I. A bridge was blown up completely losing all of its superstructure. He and the military crew repaired the bridge, with eight spans of 3 m each, using the rail line track. In the repair, firstly, the piers were rebuilt with masonry up to a certain level and then continued with a crib arrangement using timber sleepers. Finally, a built-up beam was formed to cross the span using the steel rails from the line. The beam was made up from the rails arranged in a staggered configuration and wrapped with telegraph wires. The bridge, which they repaired in one night, was bombed again the next day.

The railway constructions created an industry, market and expertise, also introducing the public to an outcome of the industrial age. The total railway lines constructed through concessions were 8343 km, and the lines within Turkey's borders were 4112 km in 1918.[47]

Road Projects: In Anatolia, the Romans had used wheeled transport, whereas the Ottomans favoured caravans of pack animals using camels, especially for long-distance trade. Camels were 20% cheaper than wheeled travel and had some advantages for long-distance travel as camels were highly resistant to hunger and droughts.[48] While camel was favoured, the wheel transport was still used and cart use was advanced with diverse types in Anatolia. One of the very common types was the tumbrel (kağnı) which was easily manufactured by villagers and used in

552   THE ENGINEER   DEC. 3, 1920

dams." The piles were driven to a depth of 30ft. below the river bed.

Just before Turkey entered the war, nine of the spans had been completed, but the steel work for the tenth span was still lying on the wharf at Tripoli. Two days after that steel work had been removed by possession of the Baghdad Railway was, at the time this article was written, ninety-six. Of that number, 37 were in service, 18 under repair, and 41 either awaiting repair or had been damaged beyond repair, either accidentally by the exigencies of the service, or purposely by the Germans during the retreat in being 66 tons. The weight on each of the two rear coupled axles is 15,650 kilos. (say, 18.4 tons), the leading coupled axle having 50 kilos. more. The tender can carry 20 cubic metres (4400 gallons) of water and 6 tons of coal. The Westinghouse automatic brake is fitted.

FIG. 1—PROFILES OF BRIDGES

the Germans, the French bombarded the Tripoli-Homs Railway and destroyed communications.‡

The Euphrates, previous to the completion of the bridge, was crossed by the railway on a temporary wooden structure, which was used for bringing all materials to the falsework and upon which the girders were erected. The timber for the temporary bridge 1918. There were, in all, nine types of locomotive, as follows :—

Type I.: Express Compound 4-6-0, with Bogie Tender and Schmidt Superheater.—There were two engines of this type, Nos. 501 and 502—see profile in Fig. 5. They were designed for a maximum speed of 100 kiloms. (say, 62 miles) per hour, and were made Type II.: 2-6-0 Simple Engine, with Six-wheeled Tender.—Of this type there are, in all, 19, and they are all identical as to the engine, though in some, as will be pointed out, the capacities of the tenders are different. Locomotives Nos. 611 to 616 and Nos. 623 to 625, both inclusive, are identical in all respects. They were built by Borsig, of Tegel, and were put

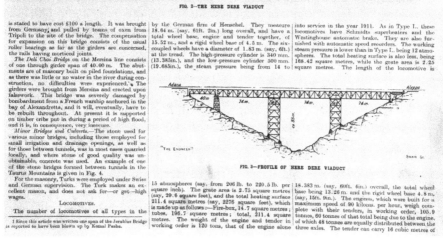

FIG. 2—THE HEBE DERE VIADUCT

is stated to have cost £100 a length. It was brought from Germany, and pulled by teams of oxen from Tripoli to the site of the bridge. The compensation for expansion on this bridge consists of the usual roller bearings as far as the girders are concerned, the rails having morticed joints.

The Deli Chai Bridge on the Mersina line consists of one through girder span of 40.00 m. The abutments are of masonry built on piled foundations, and as there was little or no water in the river during construction, no difficulties were experienced. The girders were brought from Mersina and erected upon falsework. This bridge was severely damaged by bombardment from a French warship anchored in the bay of Alexandretta, and it will, eventually, have to be rebuilt throughout. At present it is supported on timber cribs put in during a period of high flood, and it is, in consequence, very insecure.

Minor Bridges and Culverts.—The stone used for various minor bridges, including those employed for small irrigation and drainage openings, as well as for those between tunnels, was in most cases quarried locally, and where stone of good quality was unobtainable, concrete was used. An example of one of the stone bridges formed between tunnels in the Taurus Mountains is given in Fig. 4.

For the masonry, Turks were employed under Swiss and German supervision. The Turk makes an excellent mason, and does not ask for—or get—high wages.

LOCOMOTIVES.

The number of locomotives of all types in the by the German firm of Henschel. They measure 18.64 m. (say, 61ft. 2in.) long overall, and have a total wheel base, engine and tender together, of 15.52 m., and a rigid wheel base of 4.5 m. The six-coupled wheels have a diameter of 1.83 m. (say, 6ft.) at the tread. The high-pressure cylinder is 340 mm. (13.385in.), and the low-pressure cylinder 500 mm. (19.685in.), the steam pressure being from 14 to into service in the year 1911. As in Type I., these locomotives have Schmidts superheaters and the Westinghouse automatic brake. They are also furnished with automatic speed recorders. The working steam pressure is lower than in Type I., being 12 atmospheres. The total heating surface is also less, being 168.42 square metres, while the grate area is 2.25 square metres. The length of the locomotive is

FIG. 3—PROFILE OF HERE DERE VIADUCT

15 atmospheres (say, from 206 lb. to 220.5 lb. per square inch). The grate area is 2.75 square metres (say, 29.6 square feet), and the total heating surface 211.4 square metres (say, 2276 square feet), which is made up as follows :—Fire-box, 14.7 square metres ; tubes, 196.7 square metres ; total, 211.4 square metres. The weight of the engine and tender in working order is 120 tons, that of the engine alone 18.383 m. (say, 60ft. 4in.) overall, the total wheel base being 13.26 m. and the rigid wheel base 4.8 m. (say, 15ft. 9in.). The engines, which were built for a maximum speed of 90 kiloms. per hour, weigh complete with their tenders, in working order, 105.9 tonnes, 60 tonnes of that total being due to the engine, of which 48 tonnes are equally distributed between the three axles. The tender can carry 16 cubic metres of

‡ Since this article was written one span of the Jerablus Bridge is reported to have been blown up by Kemal Pasha.

Anatolia for many years. The image of a woman carrying guns with the Kağnı for the soldiers was also one of the symbols of the Independence War.

From the accounts of foreign travellers, it is known that wheeled transport was not common in Anatolia. The vehicles started to be used in cities stimulated the development of the roads. Then, the inner-city transportation needed to be designed for certain widths and geometric requirements.

The earliest road designed specifically for wheeled travel was the postal road between İstanbul and İzmit in 1834.[49] This road deteriorated quickly, however, and the transport was resumed by horse in 1840.[50]

The First Ottoman Nafia road project was the Edirne–İstanbul road in 1846. An expert was brought from Austria to investigate and determine the cost of the project. The plan was to engage local people in the road works.[51]

The next roads for wheeled transport were in Bursa Province, which was also chosen as a pilot area to practise the new regulations. These projects also started the determination of the road layout with modern engineering techniques using theodolites.[52] Construction of the 34.5 km Bursa–Gemlik road was finished in 1865 taking a total of 15 years.[53] The Bursa–Mudanya road was a 31 km strategic road; as the food supply of the capitol İstanbul was provided through the sea transport from Mudanya, it was finished in 1883.[54] Both of these road layouts were prepared by a Bursa Province chief engineer, M. Padeano, who was working in Nafia. Another four foreign engineers, working in Nafia, also took a role in the construction of the road.[55]

The Bursa–Mudanya road had a bridge called Geçit over the Nilüfer[56] River dated 1886[57] from its inscription.[58] Geçit Bridge had one span with 24 m length and 6.85 m width. The stones were placed radially along the arch and provided a neat and modern appearance to the bridge.

The surname "Leclercq" is carved on an abutment sidewall.[59] Engineer (Alfred) Leclercq was a Belgian military engineer. He worked as a teacher at engineering school and was a chief engineer for public works for the Empire. His books were translated and used as textbooks in engineering school (Figure 2.23).[60]

Mudanya road extended to İnegöl, and a bridge called Deliçay on this road was opened with a formal ceremony as known from a photograph dated 1901.[61]

Another bridge in Bursa was planned to be constructed on the Orhaneli (Atranos) road to cross over Kocasu River, dated 1912 and signed by engineer A. Voudral. The bridge would be a single-span bowstring type with a 30.4 m clear span between abutment walls.[62]

The other early road project was the 314 km Trabzon–Erzurum road constructed in 22 years (1850–1872). French engineers worked on this road, and consequently, inns along the road were named after famous hotels in Paris.[63]

These early road projects took too long and required high budgets; therefore, the empire started to look for alternative options for road-building. In 1856, a formal instruction was published for the road constructions to be carried out by locals. After 10 years of implementation, this procedure became law in 1866 as the Roads and Bridges Regulations and road tax levied on local beneficiaries. This system was similar to the derbend system of the empire and used effectively by some governors.

Ottoman Nafia programmes included road projects; however, budget restrictions were tight and did not allow for further improvements. In 1909, Regie was

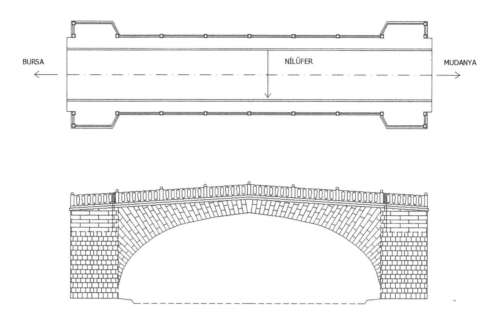

Figure 2.23 Geçit Bridge in Bursa (author).

given a concession for all new road constructions and the maintenance of the existing road networks. Regie prepared many "typical/standard"[64] projects and even imported some bridge superstructures; however, their work ceased because of World War I. Typical bridge projects were mostly masonry and timber with some occasional metal bridges using imported superstructures.

Regie Company, formed by a consortium of European banks, was the largest foreign investor in the country, and also the largest corporation. The capital of the company made up around 23% of total foreign direct investment in the Ottoman Empire in 1881–1914.[65]

In 1908, the total roads are given as 17,434 km macadam[66] road and 4450 km un-paved (soil).[67] Road construction virtually ceased due to war, and road tax could not be collected between 1914 and 1923, with projects only for military security purposes continuing. The Nafia (1933) report gives the inherited roads in 1923 as 13,885 km macadam and 4450 km soil road.[68]

Local Districts and Three Remarkable Governors: During the Ottomans and into the Republic, road- and bridge-building was managed through local authorities in the provinces called "Vilayet". Their budget and activities were partly managed and controlled by Nafia. Three governors should be mentioned in this era who made important contributions to public roads activities and bridge-building.

Ahmed Şefik Midhat Pasha (1822–1884) was one of the leading officials in the late Tanzimat period. He was appointed as governor to the newly established Danube Province called "Tuna Vilayeti" in 1861. The case was used as a "pilot" study in setting and improving the implementation of the law for the local districts. During his governorship, he showed great success by completing 3000 km of road in 2.5 years.

He engaged the local people in working as labour or donating to the works. He also established an organization called "Menafi Sandığı" (Benefit Funds), which was considered as an initiator of the Ziraat Bankası (Agriculture Bank).[69]

He then worked in Baghdad, where he constructed a 1000 km of horse-drawn tramway line. Later, he was appointed to Aydın Province, where he planned the road layouts.

Ahmed Vefik Pasha (1823–1891) was a governor of Bursa Province. He constructed many local and city routes during his period.

Halil Rıfat Pasha (1827–1901) was a governor of Sivas Province. He played a leading role in road-building and had important contributions. His saying "The place is not yours, if you cannot reach there" is still frequently used as a motto by KGM. He completed 1400 km of road in very harsh geological conditions in the region.

The Capitol (İstanbul) Projects: As the capitol city of the Ottomans, İstanbul had a leading role in engineering activities including surveying, mapping, building, sewerage and water network, tramway lines, lighting of the city, and road infrastructure and transportation projects, especially for Golden Horn and Bosporus crossings.

Some of these numerous projects were interesting. For example, tramway lines and tunnel projects were constructed to improve the city transportations. The funicular tunnel between Karaköy and Beyoğlu constructed between 1872 and 1875 by French engineer Henry Gavand still operates today.

A plan to reorganize the entire city of İstanbul was first mapped out by German military commander Helmuth von Moltke in 1839. Moltke suggested main roads be designed with a 15 m width landscape and pedestrian paths.

Moltke also envisioned both sides of the Golden Horn being cleared. Wooden piers were to be replaced by stone piers; 3 m sidewalks were to be built on either side, while trees would be planted along the road.

İzmir, as an important transportation hub on the Aegean coast, was also under development, especially for its port to be compatible with increased transportation needs.[70]

Golden Horn Crossings Proposals: The Golden Horn is the natural estuary between the "old historical city" and the opposite side of the city. In medieval times, there was no crossing and a chain was used to control the entry to Golden Horn.

The Golden Horn was a vital crossing for the city and also the country. Crossing the Golden Horn can be considered as the showcase of the engineering history of the country. It is 7.5 km long and, at its widest, 750 m across. The depth is around 40 m with another 40 m of deposits requiring foundations to need a decent depth. It is crossed at two locations: Galata is the location at the mouth of the Estuary, and Unkapanı is the industrial zone location further upstream.

Leonardo da Vinci proposed to cross the Golden Horn at Unkapanı by a stone arch bridge with 240 m span, 43 m height and 24 m width. The bridge was sketched in a letter sent to Ottoman Empire in 1502. It was not found feasible and not constructed at that time. The bridge design was brought to life again by Vebjørn Sand with a scaled-down glulam timber version footbridge constructed in Oslo with 40 m span, 5.8 m height and 2.8 m width in 2001.

Another proposal was supposedly made by Michelangelo the same year, but no further information is available about this proposed design.

An 1850 dated proposal was by prominent engineer Fairbairn, who had been to İstanbul before and requested to provide technical expertise. He described the Tubular Bridge modified with additional road and pedestrian paths. Fairbairn's letter sounds like a reply to a request, as he writes "…These bridges are now in general use for Railways, common carriage roads and foot passengers".[71]

Several proposals were made after the Earthquake in İstanbul damaged the city devastatingly in 1894; new attempts were made to reconstruct the city.

The 1894 dated proposal was by d'Antoine Corenty, who was born in Constantinople. His proposal was a 440 m long sheer sized bridge with 60 m wide. The spans were arch type with four on either side of 40 m bascule opening in the middle. Bridge had three storeys of continuous buildings on either side of the carriageway. The project is reminiscent of a colossal building of apartments rather than a bridge. The bascule opening is designed with a metal arch bridge with braced spandrel.

Another proposal in 1895 was from the MAN Company from Nurnberg. In the proposal, two options were offered. The first alternative had fixed piers in the middle with an 80 m lift span to provide passage for navigation. The rest of the bridge superstructure was truss resting on abutments. The second proposal was a revolutionary type as it suggested lifting the middle of the truss member. This option was a hybrid bridge of suspension–truss systems combined (Figure 2.24).

Joseph Antione Bouvard, head architect of the Paris Municipality, remotely designed roads and squares for the capitol. In 1902, Bouvard proposed a remarkable bridge which was a multiple span version of the Alexander III Bridge[72] in Paris. Bridge had the same features for ornaments like the same arch decorations, garlands and parapets. However, the winged horse at the top of the pillar at the entrance was replaced by a dome with a crescent moon on its top.

In 1902, another proposal was by Monsieur Pierre from Paris. The design constituted three sections; approaches were through truss with top bracings, and the

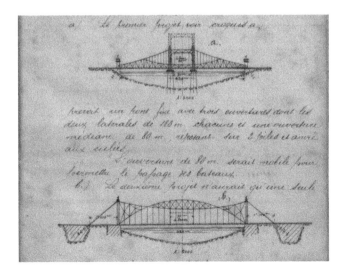

Figure 2.24 1895 dated MAN proposal with two alternatives (BOA-Y.PRK.TNF. -4-21_3).

middle was spanned with truss girders. Between the sections and at either end of the bridge, there were four sets of towers reminiscent of a minaret.

Again, in 1902, a further proposal was D'Aronco's Bridge, which had a very wide deck and also had wide platforms, stairs and landings leading to water level to reach shops, etc. These extensions nearly increased the bridge width by double.

A 1911 dated proposal for Galata crossing was for a railway bridge by Philip Holzman Company as part of the rail network within İstanbul. It was a pontoon truss-type bridge, and its height was 9 m to the deck level from water level (Figure 2.25).

Another, last but not least, proposal was an extraordinary bridge with two arch spans of 272 m with a 47 m opening in the middle for navigation. In the notes of the bridge sketch, the middle part is defined to be movable. The whole bridge used the truss-type structure for piers, arch, deck and the approach span piers. Arch spans have hinges at the crown. Four trestle towers stand at the ends of the arch spans, carrying two decks called upper and lower decks (Figure 2.26).

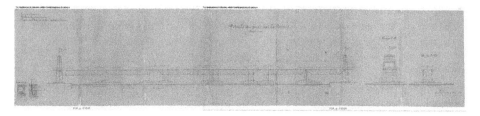

Figure 2.25 1911 dated railway bridge proposal by Philip Holzman (BOA-PLK.p.3500-4).

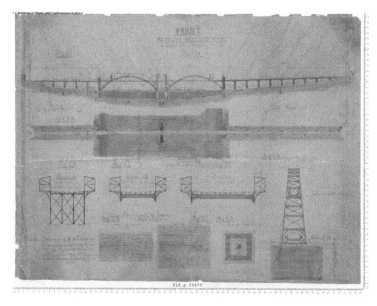

Figure 2.26 An extraordinary arch bridge with trestle towers proposal by unknown (BOA-PLK.p.6499).

Table 2.2 Golden Horn Bridge Crossings (author)

Unkapanı and Upstream Location	Galata Location
1836: Wooden Pontoon Bridge	1845: Wooden Pontoon Bridge
1863: Jewish Bridge	1863: Wooden Pontoon Bridge
1875: Iron Pontoon Bridge	**1875: Iron Pontoon Bridge**
Swing bridge operated with manpower.	Swing bridge operated by a tugboat.
52 m movable span.	Bridge total length is 480 m with 14 m
Bridge total length is 481 m with 18 m width.	width.
Cont.: Forges et Chantiers de la	Cont.: Wells and Taylor
Mediterranée	Designer: Thomas Page
Designer: -	
1912: Iron Pontoon Bridge	**1912: Iron Pontoon Bridge**
1875: Galata Bridge reused in this location	Swing bridge operated with an electric
	machine.
	67 m movable span.
	Bridge total length is 466.5 m with 25 m
	width.
	Cont.: MAN- Maschinenfabrik Augsburg
	Nürnberg A.G.
	Designer: Anton von Riepel
1940: Gazi Bridge–Steel Pontoon	**1992: Steel Bascule Bridge**
Bridge	Bascule bridge operated by hydraulic
Swing bridge is operated by a tugboat.	system.
76 m movable span.	80 m clear movable span.
Bridge total length is 454 m with 25 m width.	Bridge total 488 m with 42 m width.
Cont.: MAN- Maschinenfabrik Augsburg	Cont.: STFA – Sezai Türkeş–Feyzi Akkaya
Nürnberg A.G.	Designer: Leonhardt und Andrä, Consult.
Designer: Gaston Pigeaud	Eng.

These proposals were mostly exaggerated and unfeasible, as they aimed to impress the Sultan in the hope of a generous budget for the project. None of these proposals proceeded.

Built Bridges over Golden Horn: The bridges built over Golden Horn are given above in Table 2.2 for their designers and technical features. These bridge projects are good indications for the economic, political and social conditions within the Empire as they were prestigious projects in the heart of the capitol.

The first bridge with known records was for Unkapanı constructed in 1836. It was a wooden pontoon bridge and had an arched opening to provide navigation. Starting from 1875, bridges were made from metal and had movable parts. These bridges were all of the pontoon type. This type was chosen for both financial and technical practicality reasons, in which bridge superstructure rested on floating supports and was secured by anchors without foundations. Movable spans had been swing type opened with a small tug boat (Figure 2.27).[73]

The 1912 dated bridge built by a German Company, MAN, was the longest surviving of Golden Horn bridges. This inhabited bridge caught fire in one of its shops in the year 1992, while the current Galata Bridge was under construction. After the fire, it was left in Golden Horn for years. Later on, during its removal, some of its parts were observed to be missing (Figure 2.28).

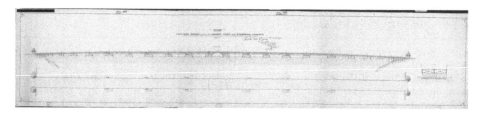

Figure 2.27 1875 dated Galata Bridge sketches, submitted together with specifications dated 18 September 1869 by George Wells. Proposal was initially for Unkapanı location (BOA PLK-p-00090).

Figure 2.28 1875 dated Galata Bridge photographed by M. Iranian (Retrieved from the Library of Congress, https://www.loc.gov).

This bridge had a great social impact on people's life in Turkey competing with the Bosporus Bridge for importance.[74] It was decorated with an arch during the visit of foreign politicians and used in movie scenes, especially for romantic settings.

A famous poem about the 1912 Galata Bridge was written by Orhan Veli Kanık, who was an influential forerunner of modern Turkish poetry.

"Hanging out on the bridge,
I watch all of you with pleasure.
Some of you pulling the oars, backwards;
Some of you picking mussel from pontoons;
Some of you clutch the helm of barges;
...
Some of you, tugboats, with funnel lowered,
Glide quickly under the Bridge;
..."

Current bridges in the Golden Horn are the 1940 Gazi Bridge in Unkapanı and the 1992 Galata Bridge in Galata locations.

The 1940 dated Gazi Bridge (MTL-07), also known as Atatürk Bridge, is again pontoon type as the World War II conditions led to tight economic conditions. This bridge, honoured with its name given by the founder of the Republic, however, is far from being of any technical or appealing speciality.

The 1992 dated Galata Bridge was built by the Turkish firm STFA, which was founded by Sezai Türkeş and Feyzi Akkaya, the prominent engineers of the early Republican period. This bridge had a fixed foundation system with piles in Golden Horn which made its construction very challenging. The bridge has a clear span of 80 m, and the structural length of the bascule span is 87.12 m which in 2008 was still the longest in the world (Figure 2.29).

Bosporus Strait Crossings Proposals: The Bosporus Strait is a channel between the Asian and European continents; therefore, the connecting bridge also connects the continents. The history of the Bosporus Bridge dates back to at least 518 BCE, when the first bridge made across the strait, to our knowledge, was constructed by Architect Mandrokles for Persian King Darius' military expeditions. This bridge was pontoon type made by connecting boats with chains or ropes.

During 1867–1900 when the railway boom occurred in Turkey, many proposals were made for crossing this famous strait. Some interesting proposals such as an Innovative Bridge by Carl von Ruppert (1867), Steel Arch Bridge by James Eads (1877), Steel Bowstring Bridge by Giano and Gourrier (1891), Transporter Bridge of Arnodin (1900) and Steel Girder Bridge with Minaret Piers (1900) will be briefly mentioned here.[75]

Ruppert's design was first exhibited at the Paris International Exhibition in 1867. The bridge had three spans with top and bottom chords curved forming a

Figure 2.29 Golden horn bridges 1940 dated Gazi Bridge, 2013 extradosed bridge and 1997 steel bascule bridge in the very front (author).

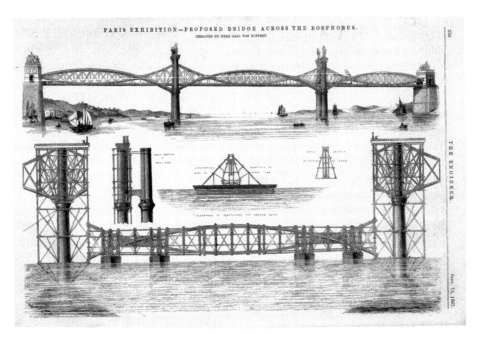

Figure 2.30 Ruppert's proposal sketch showing the construction stage of the bridge with scaffolding carried on boats, the Engineer Magazine dated 13 December 1867 (www.gracesguide.co.uk.)

so-called lenticular truss-type bridge. Continuous chords, anchored at iron piers, created an elegant appearance (Figure 2.30).[76]

Although no details are available other than his description, the proposal by James Eads Buchanan in 1875 was probably similar to his renowned Mississippi Bridge.[77] Eads' proposal for Bosporus was 15 spans with a central span of 228.6 m arch resting on granite piers, and a total bridge length of 1829 and 37 m above the water level. As understood from the information provided by Eads, he was requested for the proposal by the Grand Vizier. He prepared the plans and costs and sent them to the Empire. He had been assisted by Lambert in preparing the submission.[78]

The proposal that the two Constantinople-based Franco-Belgian engineers Giano and Gourrier presented in August 1891 was a six span bridge with iron piers and bowstring steel arches forming a continuous superstructure.

A transporter bridge proposal of Arnodin had three giant towers with a 720 m span length. He was famous for transporter bridges and had built many with a span range of 110–170 m. This bridge was intended to carry the road and pedestrian traffic across the strait on a platform hanging from the main boom. This bridge would have had minarets on top of the towers and mosque-like structures for the main cable anchorages (Figure 2.31).

Another proposal was a railway bridge from the historical peninsula directly to the Anatolian side crossing Bosporus. This bridge had amazing features such as

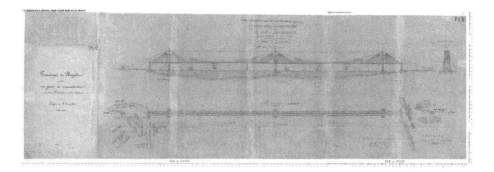

Figure 2.31 Arnodin's proposal for a transporter bridge (BOA-PLK.p..3500).

Figure 2.32 An interesting bridge with mosque-like piers proposed by unknown (BOA-
PLK.p..3500).

mosque on top of the piers with domes adorned by colourful bricks, tiles and glit-
tering metals. Piers were described to have "open places with rotating self-defence
system" (Figure 2.32).[79]

Sadly, none of these projects was realized and the Bosporus remained uncon-
quered for many years.

2.5   END OF THE EMPIRE

Despite efforts in catching up with worldwide standards, especially in a technological manner, these projects were indeed all run by foreign investors. The decline of the Ottoman Empire continued until World War I. The Ottoman Empire entered World War I in 1914 in an alliance with Germany and emerged defeated from the war in 1918, compelled to sign the Mudros Armistice on 30 October 1918. Under the terms of this Armistice, the territories of the Ottoman Empire were occupied.

In response to the decline and occupation of the empire, a national resistance emerged and was mobilized under the leadership of Mustafa Kemal Atatürk, a military commander and field marshall in 1921. This independence struggle was carried out in parallel paths of diplomacy and on the military field. A new parliament was established in Ankara in 1920, while the Battle of the Sakarya (Sangarios) was still driving only 80 km far from Ankara. Ankara later became the capitol of the Republic.

The national struggle against occupying forces ended in 1922, and the Foundation of the Republic was officially proclaimed on 29 October 1923. The Ottoman Sultanate, also the Caliphate of Muslims, was abolished in 1924 with a parliamentary decree. As a result, Turkey became a secular state with her people as citizens of the Republic rather than the subjects of the Sultanate or the slaves of colonization. This new system in which the people are citizens and the government exists to provide public services was a true revolution.

NOTES

1. https://whc.unesco.org/en/list/1572
2. Godley A. D. (2010) *Translation of "The Histories" of Herodotus of Halicarnassus*. Pax Librorum republished after 1920.
3. Britannica, The Editors of Encyclopaedia. "Fertile Crescent". Encyclopaedia Britannica, 7 Apr. 2020. Retrieved from: https://www.britannica.com/place/Fertile-Crescent.
4. For bridges mentioned in the text, please refer to blog: https://kantaratlas.blogspot.com.
   The heritage stone bridges in Turkey have been studied by Çulpan, İlter and Tunç. Çulpan intends to cover all bridges in Anatolia and in Balkans, and 135 bridges in Turkey are extensively described in the book. İlter studied the bridges before Ottomans, covering 46 bridges in detail. Tunç's book is an inventory for stone bridges with 260 bridges in total.
   Çulpan, C. (1975) *Türk Taş Köprüleri*. Türk Tarih Kurumu. Ankara.
   İlter, F. (1978) *Osmanlılara kadar Anadolu Türk Köprüleri*. General Directorate of Highways. Ankara.
   Tunç, G. (1978) *Taş Köprülerimiz*. Karayolları Genel Müdürlüğü. Ankara.
5. Clarke, J. T., Bacon, F. H., and Koldewey, R. (1881–1883) Investigations at Assos: Expedition of the Archaeological Institute of America. Part I–V.
6. Guinness World Records. (2017). Retrieved from: http://www.guinnessworldrecords.com/world-records/oldest-bridge.
7. Tyrrell, H. G. (1911). *History of Bridge Engineering*. Tyrell. Chicago.
8. Main references for Roman bridges in Turkey:
   Galliazzo, V., and Chevallier, R. (1994). *I ponti romani*. Canova. Treviso.
   O'Connor, C. (1993) *Roman Bridges*. Cambridge University Press. Cambridge.
   Comfort, A. (2014) Roman bridges of South-East Anatolia. In Bru, H. (Ed.), *L'Anatolie des peuples, des cités et des cultures*. Vol. 2. Besancon. Pages: 315–342.
   Wiegand, T. (1904) Reisen in Mysien. In. *Mitteilungen des Deutschen Archäologischen Instituts, Athenische Abteilung*, Vol. 29, Berlin. Pages: 254–339.

9. On its eastern, downstream side, a Christian inscription in Greek runs along most of its length, citing almost verbatim Psalm 121, verse 8 of the Bible. The text reads:
Kýrios ho Theós phyláxei tēn eisodón sou ke tēn exodón sou apó tou nyn kai héōs tou aiōnos, amén, amén, amén. [The] Lord God may guard your entrance and your exit from now and unto all time, amen, amen, amen.
Doomed by the Dam (1967) A Survey of the Monuments threatened by the Creation of the Keban Dam Flood Area. Elazıg, 18–29 October 1966, 9, METU. Faculty of Architecture. Pages: 54–57.
10. Herodian of Antioch (1961) *History of the Roman Empire.* Translated by Edward C. Echols. University of California Press. USA.
11. Choiseul-Gouffier, A. (1823) *Voyage pittoresque de la Grèce.* Paris. Page: 307.
12. Bayraktar, S. (2018b) Trabzon Dernekpazarı Yenice Köyü'nde Tarihi Ahşap Bir Köprü. Süleyman Demirel Üniversitesi. Fen-Edebiyat Fakültesi Sosyal Bilimler Dergisi. No.44.
13. Derbend is a Persian word derived from der: pass and bent: holding.
Orhonlu, C. (1967) *Osmanlı İmparatorluğunda Derbend Teşkilâtı.* Üniversitesi Edebiyat Fakültesi. İstanbul.
14. https://whc.unesco.org/en/tentativelists/6042/
15. Çulpan (1975) Page: 190.
16. Turan, Ş. (1963) Osmanlı Teşkilatında Hassa Mimarları. *Tarih Araştırmaları Dergisi,* 1:1, 157–202.
17. Military and education were the earliest areas where changes occurred; for example, the Corps of salaried bombardiers (ulufeli humbaracı ocağı) was founded in 1735 as a first institution that introduced systematic techno-military training and later schools opened in artillery and mathematical in 1772 and 1775, respectively. The mathematical school was later named "engineering school" in 1781. In 1795, a new engineering school was established to train engineers and artillery officers. After 1806, this school was named Imperial Land Engineering School (Mühendishâne-i Berrî-i Hümâyûn).
18. The term used for public works was "imar", which means civilization and development excluding the social content. The term later changed to "Nafia", which can be translated as utility, to refer to public assembly at the governance level. Nafia also later changed to "Bayındırlık", which means prosperous, in 1935.
19. Even though conversion to İslam was needed for non-Muslims to work during the Ottoman reign, this tradition was out ruled since Baron de Tott.
20. Martykanova, D., and Kocaman, M. (2018). A land of opportunities: Foreign engineers in the Ottoman Empire. In Roldán, C., Brauer, D. and Rohbeck, J. (Eds.), *Philosophy of Globalisation.* De Gruyter. Berlin. Page: 244.
21. Atam, Ş. (2015) Osmanlı Devleti'nde Nafia Nezareti, P. Hd. Thesis submitted to Niğde University. Page: 193.
22. The word kaldırım in Turkish derived from the root kaldır: lift may refer to road surface to be raised for drainage purposes.
23. Dinçer, C. (1972) *Osmanlı Vezirlerinden Hasan Fehmi Paşa'nın Anadolu'nun Bayındırlık İşlerine Dair Hazırladığı Layiha.* Vols V–VIII. T.T.K. Belgeler Dergisi. Ankara.
24. Geyikdağı, N. (2011) *Foreign Investment in the Ottoman Empire: International Trade and Relations (1854–1914).* Library of Ottoman Studies 27. Tauris Academic Studies. New York. Pages: 140–141.
25. Tekeli, İ., and İlkin S. (2004) *Cumhuriyetin Harcı: Modernitenin Altyapısı Oluşurken.* Bilgi Üniversitesi Yayınları. İstanbul.
26. Martykanova, D. (2010) *Reconstructing Ottoman Engineers: Archaeology of a Profession (1789–1914).* Edizioni Plus. Pisa.
27. CUP (Committee of Union and Progress) İttihad ve Terakki Cemiyeti in Turkish.
28. Young Turks (Turkish: Jön Türkler or Genç Türkler) was a political reform movement in the early 20th century that favoured the replacement of the Ottoman Empire's absolute monarchy with a constitutional government.
29. Shaw, S., and Shaw, E. K. (1977) *History of the Ottoman Empire and Modern Turkey, Vol. 2, Reform, Revolution and Republic: The Rise of Modern Turkey. 1808–1975.* Cambridge University Press.UK.

30. **Francis Rawdon Chesney** (1789–1872), a British general and explorer, is usually thought of as a pioneer of canalization.
31. Chesney, F. R. (1857) *Report on the Euphrates Valley Railway*. Smith, Elder. London.
32. The eminent civil engineer Robert Stephenson (1803–1859) surveyed and built his line.
    Bailey, Michael (2003) *Robert Stephenson – The Eminent Engineer*. Ashgate. London. Pages: 149–158.
33. 319 km İstanbul- Edirne and 149km Dedeağaç-Edirne completed in 1873.
    Yıldırım, İ. (2001) *Cumhuriyet Döneminde Demiryolları (1923–1950)*. Atatürk Araştırma Merkezi Yayınları. Ankara. ISBN:975-16-1477-5
34. Engin (1993) mentioned that three bridges collapsed with an intensive rain during the investigations of the commission, and another information is that the bridges were constructed as per the contract which has foreseen the use of timber.
    Hertner (2006) mentions: "By then (1881) also most of the original 257 wooden bridges used by the Oriental Railways had been substituted by iron ones".
    Engin, V. (1993) *Rumeli Demiryolları*. Eren Publication. İstanbul. Pages: 171–175.
    Hertner, P. (2006) The Balkan Railways, International Capital and Banking from the 19th Century until the Outbreak of the First World War, Bulgarian National Bank Discussion Papers. Page: 16.
35. **Thomas Page** (1803–1877) Prominent bridge engineer famous for the design of Westminster Bridge and first Chelsea Bridge both crossing over Thames River in London. He is also a designer of 1875 Galata Bridge.
    BOA. HR_SFR_3_00058_00008_001_001 is a letter from Tomas Page, which is about an accident in which his son got shut in his arm and comments about the route of the line.
    BOA. HR.SFR.3..._46-6_3 is 1859 letter that Thomas Page had been advised by Fairbairn.
    BOA.I.. HR..181–10066_3 is 1860 dated letter Thomas Page requesting a concession for Golden Horn Bridge.
36. Özyüksel, M. (1988). *Anadolu ve Bağdat Demiryolları*. Arba Publication. İstanbul. Page: 15.
37. **Wilhelm von Pressel** (1821–1902) Austrian railway engineer or technician. He is also considered as father of Baghdad railway for his personal efforts to make relations between Germany and Empire and supplying finance for the project. Later: "the father of the railway, was mysteriously absent from the rash negotiations taking place in late 1888". He died in 1902 in İstanbul.
    McMurray, J. S. (2001) *Distant Ties: Germany, the Ottoman Empire, and the Construction of the Baghdad Railway*. Praeger. Westport, Conn. Page: 23.
38. BOA State Ottoman Archives, HRT-373: Reseau ferre de l'asie Mineure (Asia Minor Railway Lines).
39. Ankara–Kayseri line was not built even though it was planned, since Russia interposed serious objections to Germany not to extend the line beyond Ankara.
    Earle, E.M. (1923) *Turkey, the Great Powers, and the Baghdad Railway: A Study in Imperialism*. The Macmillan Company. New York. Page: 149.
40. Manas Efendi was also mentioned as an engineer who stayed during Independence War and repaired bridges.
    Gürel, Z. (1988) *Kurtuluş Savaşı'nda Demiryolculuk*. VIII. Belleten, C. LII, S.205.
41. Revue Technnique d'Orient (1912) Le Sakaria, Ponts anciens et pont modernes. İstanbul.
42. Heslop, G. (1920) The Baghdad Railway. The Engineer. Obtained from Grace's Guide to British Industrial History at https://www.gracesguide.co.uk/
    Article was written by Major Derwent Gordon Heslop, RE, Late Military Control Officer of the Maintenance and Locomotive Department of the Baghdad Railway. Published in I, II and III sections on 12 November, 26 November 1920 and 3 December 1920 dated magazines.
    In the article dated 3 December 1920, an endnote is: "Since this article was written the Seihun Bridge is reported to have been blown up by Kemal Pasha".
43. Atatürk, M.K. (1995) Nutuk. Kültür Bakanlığı. Ankara.
44. **Heinrich August Meissner** (1832–1940) Chief engineer of the construction of two lines, Hijaz and Anatolian–Baghdad lines from Germany. He spoke Turkish fluently. After completing the Hijaz line, he was awarded the title of pasha in 1904 by Abdülhamit II. In 1918, Meissner

returned to Germany, but he returned back in 1924 as adviser on building and maintenance of railroads in new republic by the invitation of Atatürk.

45. **Ahmet Muhtar Çilli** (1871–1958) Graduated from İstanbul Engineer School and worked in Hedjaz Railway. He prepared report for the plan and profiles of the Hejaz railway. He was later appointed as Meissner's assistant, and he successfully accomplished this task. He also managed the exclusively Muslim staff for the sacred section of the route.

46. **Mustafa Şevki Atayman** (1872–1958) He graduated from Hendese-i Mülkiye in 1897 and was appointed to Ankara and Kosova as an engineer. He worked in Hedjaz Railway between 1913 and 1918. After war, he worked in railway projects. He wrote his memoirs in a book.

   Atayman, M. S. (1967) *Bir İnşaat Mühendisinin Anıları, 1897–1918*. Baha Matbaası. İstanbul.

47. In the book, the numbers given by Yıldırım (2001) are used.

   The total railway line stayed within the current borders of Turkey with the collapse of Empire is also provided in other studies as 4083, 4130, 3974, 4138 km, from Nafia (1933), Tekeli, İ. and İlkin S. (2004) Page: 274, Bayındırlık (1973) Page: 46 and Engin (1993) Page: 222 in the order.

48. Tekeli, İ., and İlkin S. (2004) Page: 77.

49. Road project prepared by Hassa Müdür Ahmet Fevzi Pasha, road was 12.5 m wide.

   Çetin, E. (2017) Tanzimat'tan II. Meşrutiyet'e Anadolu'da Karayolu Ulaşımı, Türk Tarih Kurumu Yayınları, Ankara. Page: 159.

50. Tekeli, İ. and İlkin S. (2004) Page: 100.

51. Atam (2015) Page: 197.

52. Tekeli, İ. and İlkin S. (2004) Page: 109–110.

53. Çetin (2017).

54. There was already a road on this route and was converted to macadam. Therefore, resources usually mention repair of the road rather than the new construction.

   Baykal, K. (1950). *Bursa ve Anıtları*. Aysan Matbaası. Bursa.

55. The projects were prepared by M.Padeano and carried out by other foreign engineers M.M. de Leffe, Tassy, Ritter, col. Gordon Heslop. Ref: Tekeli, İ. and İlkin S. (2004) Page: 110.

   Béatrice. S.L. (1994) Ottoman Power and Westernization: The Architecture and Urban Development of Nineteenth and Early Twentieth Century Bursa. *Anatolia Moderna*. 5:1, Page: 199–232.

   Vital, C. (1894) *La Turquie d'Asie*. Paris: E. Leroux. Page: 85.

56. Nilüfer, the name of wife of Sultan Orhan, was also the name of the first bridge constructed by Ottomans on the same river, 1.5 km west to the Geçit Bridge on the old road.

57. The date of the bridge is later than the completion of the road. Bridge was probably constructed after the 1855 earthquake took place in the region, causing collapse of nine bridges. Çetin (2017) Page: 168.

58. The inscriptions were in the middle of the bridge opposite to each other. Both were made of marble with the same dimensions and details. They had the tughra of Abdülhamit on top. Inscriptions are now preserved in Turkish Islamic Art Museum in Bursa. A poem is carved on the stone, and one of the rows of the poem also gives the date of the bridge in a traditional "abjad" which gives the date with the corresponding numbers of consonant letters in the text.

59. Tunç (1978) Provides the photograph of an abutment sidewall with "Leclercq" carved on.

60. Günergün, F. (2005) Mektebi Harbiye'de okutulan mimarlık ve inşaat bilgisi dersleri için 1870'li yıllarda yazılmış üç kitap. In Mazlum, D. & Cephanecigil, G. (Eds.), *Afife Batur'a Armağan: Mimarlık ve sanat tarihi yazıları*. İstanbul. Pages: 151–163.

61. Özendeş, E. (2000) Osmanlı'nın ilk başkenti Bursa. Geçmişten Fotoğraflar. Yapı Endüstri Merkezi Yayınları. İstanbul. There is no further information found about the bridge.

62. BOA PLK-p-02082-0001 Bursa-Atranos yolunda Kocasu Nehri üzerine yapılacak betonarme köprü projesi ve resmi. (Fr.)

63. Deyrolle, T. (1939) 1869'da Trazon'dan Erzurum'a. Turkish translation by: Reşad Ekrem Koçu. İstanbul.

64. A typical project in engineering is the preparation of a generic project to use at different locations after revising for local conditions. Standard project is also used for the project

prepared for the parts of the bridges which does not change for local conditions, like parapets and joint details.

65. Birdal, M. (2010). *The Political Economy of Ottoman Public Debt, Insolvency and European Financial Control in the Late Nineteenth Century.* Tauris Academic Studies. London.

66. **Macadam** is a type of road construction, pioneered by Scottish engineer John Loudon McAdam around 1820, in which single-sized crushed stone layers of small angular stones are placed in shallow lifts and compacted thoroughly (https://en.wikipedia.org/wiki/Macadam).

67. The number is given for the 1908 borders of the empire in Noradounghian programme.

   Örmecioğlu (2010) Page: 291 and Çetin (2014) Page: 267 repeats the same numbers. The road kilometres can also be found from: Atam (2015) Page: 206 and Tekeli (2004) Page: 182 for the roads in the year 1889.

   Örmecioğlu, H. T. (2010) Technology, Engineering, and Modernity in Turkey: The Case of Road Bridges between 1850 and 1960. Doctorate thesis submitted to Middle East Technical University (METU).

68. 15. Yıl Kitabı (1938) *Cumhuriyet Matbaası.* Ankara. Page: 260. Retrieved from: http://hdl.handle.net/11543/553.

69. Atam, Ş. (2015) Page: 138.

70. Geyikdağı, N. (2011).

71. Tubes referred to Conway (1849) and Britannia (1850) bridges built with an innovative wrought iron tubes section, which were made with metal plates riveted together to form box. These bridges carry railway lines with both 122 and 152 m maximum span lengths. During the design, William Fairbairn (1789–1874) was in the team, reviewing and supervising the work together with Robert Stephenson.

   The bridge proposed has widened cross section with road and pedestrian access. The widening is achieved by cantilevered extensions as understood from the letter.

   BOA Archive letter was written by Fairbairn to Sultan dated 1850. It mentioned an attachment but was failed to find in the archives. The part of the original letter is read as: "It has further suggested modifications of the original design in its application to spans varying from 50 up to 300 ft [app. 15–90 m], which the Tubes from the parapets of the Bridge, and by the introduction of … leavers, the roading was between the Tubes and is open to the atmosphere on all sides".

   As the bridge was offered for 500 ft (152 m) in the letter with modifications to original bridge. The proposal is classified for Golden Horn crossing.

   BOA Archive letter: HR.TO.213–29_3 Ferbrek nam İngiltere'li mühendis tarafından tasarlanan demir boru köprünün modeline dairdir.

72. Alexander III Bridge was built (1896–1900) as single-span steel arch bridge with three hinged ribs. Bridge is in the Beaux-Arts style bridge, with its exuberant Art Nouveau lamps, cherubs, nymphs and winged horses at either end. The design, by the architects Joseph Cassien-Bernard and Gaston Cousin. The bridge was built by the engineers Jean Résal and Amédée Alby. It was inaugurated in 1900 for the World's Fair.

73. İlter, İ. (1973) *Boğaz ve Haliç Geçişlerinin Tarihçesi.* Karayolları Genel Müdürlüğü Publication. Ankara.

74. These two bridges are of different nature in terms of social and technical significance. Bosporus Bridge had a big impact for its size, high technology and strategic value. Galata Bridge was in everyday life, blended in and became a part of the scene.

75. Voorman, F. (2015) Crossing the Bosphorus: Nicht realisierte Infrastrukturprojekte in Spätosmanischer Zeit. *Architectura Journal*, 45:2, 145–162, DOI: 10.1515/ATC-2015-0013.

76. The Engineer (1867, 13 Dec) Magazine obtained from Grace's Guide to British Industrial History at https://www.gracesguide.co.uk/

   Ruppert combined the system of an arch bridge with that of a suspension bridge and thus referred to two important and well-known parabolic girder bridges, the Saltash Bridge built by Isambard Kingdom Brunel (aka. Royal Albert Bridge, inaugurated in 1859) and Heinrich Gerber's South Bridge over the Rhine Mainz (also completed in 1859). The main spans of these bridges, however, at 140 and 100 m, respectively.

77. Mississippi (Eads) Bridge is a combined road and railway bridge over the Mississippi River connecting the cities of St. Louis, Missouri and East St. Louis, Illinois. The bridge is named for its designer and builder, James Buchanan Eads. Work on the bridge began in 1867, and it was completed in 1874. Eads Bridge was the first bridge across the Mississippi south of the Missouri River. https://en.wikipedia.org/wiki/Eads_Bridge
78. The Bosphorus bridge (1877 September 7) Railroad Gazette. Page: 913 and the Bosphorus bridge (1877 September 14) Railroad Gazette. Page: 937.
79. Even though this drawing is repeated for Arnodin in many studies, there is no indication that it belongs to Arnodin and it is very different in style and format than the drawings belonging to Arnodin by his signature.

Chapter 3

Republic

3.1 INTRODUCTION

After the foundation of the Republic and its official establishment in Ankara in 1923, Public Works Authority/Nafia[1] resumed its duties under the newly established government.

The new Republican system was based on the principles of parliamentary democracy, human rights, national sovereignty, private ownership and secularism, and the separation of religion and state affairs. A secular education system was established, the Arabic alphabet was changed into a Latin alphabet, and new civil and criminal codes were adopted. Gazi Mustafa Kemal (Atatürk) was elected as the first president of the Republic of Turkey. Then, another war started, this time against ignorance and backwardness.[2]

In the aftermath of the war, many public buildings, especially mosques and bridges, had been damaged during the withdrawal of allied forces. For example, an organized fire in İzmir, Manisa and the environs had destroyed most of the city. As a result of the massive changes like war, deportation, immigration and population exchanges, at the end of 1924, the population of Turkey stood at around 13 million.[3]

The conditions were extremely challenging to resurrect the country after decades of neglect and years of war with the consequent scarcity of resources and machinery, undeveloped industry and the lack of expertise. In addition, the Ottoman foreign debt was renegotiated and apportioned to all of the empire's successor states. The Turkish government assumed 67% of the total, to be paid in sterling beginning in 1929.[4] Furthermore, many enterprises, especially railway lines, were still foreign-owned.

On the other hand, the people, now citizens of the Republic, were highly motivated, proud and enthusiastic to rebuild their country. First, the damages of war had to be repaired and then new buildings, infrastructure and a modern transportation network to be constructed. Similarly, bridges had to be built for this network, as the product of the same common consciousness and spirit. Therefore, this era also witnessed the "first of their types and new methods", and many bridges from the period have significant historical and social value as well as important technological features (Figure 3.1).

Most public works had previously been carried out through foreign-invested projects, which were mostly railways. They were mainly shaped for the profit of the investors and their focus had naturally been on short-term profit. Therefore, these projects had not provided much opportunity to set up a systematic technical background and organization in the country. On the other hand, industrialization and basic infrastructure

DOI: 10.1201/9781003175278-3

Figure 3.1 Covers from the 1939 Nafia Magazine published periodically describing the works, progress, activities and organization (İBB Atatürk Library).

inevitably started with the railways, which was also an introduction to the local people of the new opportunities and alternatives in their life.

The 1923 İzmir Economic Congress, held some months before the Lausanne agreement, reaffirmed the desire for economic sovereignty. As commented by Feroz[5], "Turkey demonstrated that its political leaders and the various economic groups were united around the goal of an independent national economy". The approach to concessions was confirmative as long as local rules were obeyed, and it was beneficial for the country.[6]

3.2 REPUBLICAN PUBLIC WORKS

The new Ankara government named its public works ministry "Nafia Vekaleti" on 23 April 1920, in response to the "Nafia Nezareti" of the İstanbul government. Although Vekalet and Nezaret[7] have different meanings, they both refer to "Public Works Authority". The managerial system was similar. According to Timur,[8] the change was the power passed from the Ottoman government, which was open to foreign influences, to the Ankara government, which was based on military-civil intellectuals and Anatolian notables of petty-bourgeois origin.

As given in a 1933 dated report,[9] Nafia organization was formed of the minister together with an Undersecretary, a Construction Supervision Department and four general directors for concession works, waterworks, railways and roads departments. The subsection related to roads was named "General Directorate of Roads and Bridges". The organization had administrative units such as human resources, law unit and also a technical school that aimed to educate technicians. At the same time, Nafia had establishments of State Railways, Engineering School and postal services units (Figure 3.2).

Figure 3.2 **06 March 1933-dated photo taken in Antalya with 12 Nafia engineers: Zühtü, Feyzi Akkaya, Siyami, Necmi, Besim, Mitat Kasim, Nihat, Bodos Kemal, Halil Bey, M. Demayo, Sadık Diri and Rıza Bey (FATEV archive).**

There was a chief engineer (başmühendis) for every province in charge of the supervision of the public works, another engineer as required and middle-ranking technical staff (fen memuru or kondoktor).

The Engineering School had its own budget and regulation; however, it operated under Nafia direction. Nafia also had compulsory work for graduates to work in the authority for a certain time. For example, Akkaya had to pay out the remaining amount in cash when he wanted to work in SAFERHA after his military service.

The first Nafia Minister of the Republic was İsmail Fazıl Cebesoy, who had been a soldier and parliamentarian of the empire and moved to Ankara when İstanbul was invaded and its parliament was shut down. In the following years, ministers were almost always soldiers with a few exceptions. The longest-lasting in the position were Feyzi Pirinçcioglu, Behiç Erkin, Recep Peker and Hilmi Duran with 2 years average, followed by Ali Çetinkaya[10] and Ali Fuat Cebesoy[11] who each served 5 years as ministers. Ahmet Muhtar Çilli and Süleyman Sırrı Aral were the only engineers to become ministers.

The scope of the Nafia works, as listed in their programmes, included ports, breakwaters, roads and railway constructions, swamp drainage, flood control, dams, irrigation and drainage channels and river transportation. In the early Republic era, the management of existing infrastructure assets and auditing and supervision of foreign concession projects were added to the programme. Nafia's work content and priorities changed according to circumstances. For example, after the independence war, the priority was swamp drainage to control malaria, as more than 50% of the population was suffering from it.

In the meantime, there was a significant amount of public works to be carried out throughout the country, for example, factories, hospitals, housing and public buildings with a high demand for resources. These works were carried out by ministries, provinces and municipalities, and at the same time, the private sector evolved accordingly. Republic encouraged the development of the private sector and the formation of a bourgeois class to sustain the development.

Nafia continued to prepare programmes for the public works. The first Republican Nafia programme[12] was prepared sometime before Ankara had been chosen as the capitol in 1923. The plan was a repetition of previous Ottoman plans. Foreign concessions on a project basis were presumed for carrying out the works.

The next programme[13] was in 1929 and was prepared during the ministry of Recep Peker. This programme can be considered as the main programme applied in the early Republican era, as it was prepared based on the new regime of the country and was used until 1945. The programme included diverse activities ranging from dams, ports and schools to community centre buildings (Halkevi), entertainment parks and convention centres.

In terms of transportation infrastructure, the essential focus of the programme was railway lines, and the motto was "weaving the whole country with iron webs" repeated by Prime Minister, İsmet İnönü. Road infrastructure was feeding the railway lines; however, considerable road projects and strategic bridges were also covered in the plan. The importance of the road was emphasized by Atatürk, as the villagers asked for school and road, wherever he visited and also describing the roads as "the wings of the villagers" during his speech in the groundbreaking ceremony of the Samsun Railway.

3.3   REPUBLICAN RAILWAYS

During the independence war, railway lines located in central and southern Anatolia, which belonged to Anatolian and Baghdad Railway Company, were commandeered by the military for logistic use. The railway operations were managed by Behiç Erkin (who is considered the "father of railways" in Turkey), who was put in charge of these railway operations at this time. Lines were often damaged and received temporary repairs. The resources were very scarce, and the experts to repair the bridges had to come from the Baghdad Railway line which was still under construction. And, on some occasions, existing rails on the lines were taken to re-use in new narrow gauge track lines.

After the war and the Mudanya Armistice in 1922, the public debt commission returned lines to their investors/owners. Then, in Lausanne, this agreement was signed, and the management of existing lines was handed to investors. The Republic had to buy the existing railway lines from foreign investors under the process, called nationalization, and had to pay kilometre guarantees until their nationalization (Figure 3.3).

The policy of the Republic was "one more span of railway line"[14] implying the focus on constructing new lines, and at the same time, existing lines need to be purchased especially for strategic reasons. In 1924, a debate on the nationalization of the Anatolian

Figure 3.3 1893-dated Pekdemir Viaduct on Baghdad Railway, photograph taken during the replacement of the superstructure in 1931 (FATEV archive).

Railway line resulted in the then Nafia Minister, Muhtar Çilli, to resign only 3 months after his assignment to the position.

The worldwide economic crisis in 1929 put a temporary hold on the construction of new lines as the metal could not be imported; however, this also resulted in the existing lines becoming more affordable to nationalize (Figure 3.4).

The existing lines were not interconnected, and the lines were operated by foreign capital with extremely high tariffs. For example, in 1924,

> the cost of transporting one tonne of wheat from central Anatolia to İstanbul was $8.8 whereas it was only $5 from New York to İstanbul; hence it seemed more rational to feed the population of İstanbul from Iowa than Ankara and Konya and to let the Anatolian peasant vegetate in subsistence farming.[15]

Railways must have meant more than their function for the public; the only transportation to rural places, especially in the eastern region, was provided with a railway. For example, a "Health Wagon" was used since 1930, as the last wagon of the train, which was designed to provide public health service and carry out regular checks on workers and general health and sanity issues around the city centres.

The first new railway line of the Republic era, also the first built by a local investor, was the 36-km Samsun-Çarşamba line financed by Nemlioğlu family.[16] Construction started with Mustafa Kemal Atatürk ceremonially breaking the soil with a pickaxe on 21 September 1924.

Ankara-Kayseri-Sivas line was the most strategic line since it was the connection of the capitol to the east. The first 80 km of this line had been started by military forces in 1914. The line was used efficiently during the war with narrow gauge extensions. After the war, it was completed by Sevki Niyazi Dağdelen until Yerköy station. The remaining line was completed by Emin Sazak[17] to Kayseri in 1927 and then Sivas in 1930.

CARTE DU RESEAU FERROVIAIRE DE LA TURQUIE

Figure 3.4 Railway map from the year 1935 from La Turquie Kemaliste Magazine (İBB Atatürk Library) The legend shows from top to bottom: 1. State-owned, 2. Built under the Republic, 3. Under construction, 4. Owned by foreign companies and 5. Owned by the State and operated by concessionary companies.

Figure 3.5 A temporary timber bridge at the Kızılırmak river crossing of the Ankara-Sivas line (İBB Atatürk Library).

These lines were probably already investigated and designed. The bridges on these lines were very similar to concession lines in design and material. The bridges were masonry arch bridges and metal superstructures on masonry substructures. Ankara-Kayseri's line had similar structures; however, Kızılırmak (Halys) and Delice crossings had noteworthy truss bridges with 50-m spans. The crossings at 586 km over Halys were made with a 50-m span timber trestle bridge, probably as a replacement of metal bridge design which was scarce due to global crisis. The remaining bridges were masonry bridges (Figure 3.5).

Samsun-Sivas line, which was started in 1910 by Nafia engineers,[18] was sent to the region for investigations, and then, the line was given to Regie[19] to construct. This line can be a representative of all different alternatives experienced for the construction of railway lines during the early Republican era.[20] Regie held with war, the military was, in turn, to initiate the construction during the war, but they could not complete it either. Afterwards, foreign investors, Chester[21] and company SIT,[22] were given the line to construct; however, this attempt as well failed. The line was finally completed in 1932 by Nafia engaging local firms. One of them was Nuri Demirağ,[23] who has been given this surname, meaning "ironnetwork", by Atatürk. Samsun-Sivas line had many steel bridges with spans 10–30 m all in truss type, and the longest span was Kızılırmak crossing with 50-m-long underspan truss bridges (Figure 3.6).[24]

Then, the two lines, Fevzipaşa-Diyarbakır (also called the copper line) and Irmak-Filyos (also called the coal line), were awarded to NOHAB.[25] Nohab prepared the design and managed and supervised the construction activities using local subcontractors. The report prepared by the company provides useful information about the engineering of the lines and the conditions. One of the challenges faced constructing in Turkey was defined as the scarcity of roads to transport the construction materials, the other was the lack of expert workers, and therefore, many foreign engineers were used at the beginning of the project; however, over a short time, the locals gained expertise.[26]

These two lines had diverse and noteworthy bridges, like the reinforced concrete arch, masonry arches with pierced spandrels and bowstring concrete bridges.

The copper line, Fevzipaşa-Diyarbakır, had Göksu and Fırat bridges, both crossing over Fırat river. Göksu is an arch bridge with seven spans each with 35 m and a total length of 294 m completed in 1929. The spans are open spandrel with an arch of crescent shape. 1932-dated Fırat bridge is an eight span segmental arch with 55 m lengths for deck arches and 25 m for solid spandrel arches (Figure 3.7).

Construction of these bridges was carried out with three hinged metal scaffolding arches as timber was scarce in the region. For the Fırat bridge, the metal

Figure 3.6 **One of the underspan bridges of the Samsun-Sivas line (SALT research, miscellaneous photographs and postcards).**

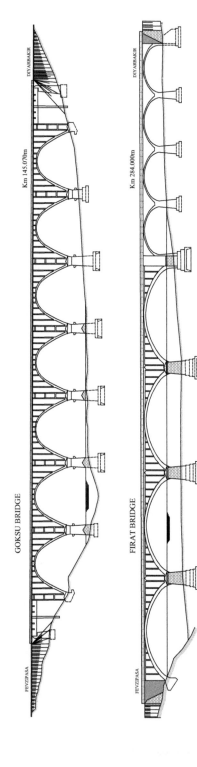

Figure 3.7 Göksu and Fırat bridges of Fevzipaşa-Diyarbakır Railway line from NOHAB album (author).

Figure 3.8 Construction of Fırat bridge, with the temporary timber trestle bridge in the front, and the scaffolding supports are still present for the arch (SALT research, miscellaneous photographs and postcards).

scaffolding was completed with timber for its upper elevations for the desired geometry (Figure 3.8).

The coal line, Irmak-Filyos, had a Soğanlı bowstring concrete bridge with a 51 m span and significant segmental arch bridges. Masonry bridges, like Devres bridge with 32 m spans, are very appealing with their skewed layouts, pierced piers and stones perfectly cut with smooth finishes (Figures 3.9 and 3.10).

Sivas-Erzurum line, called the iron line, was the first line awarded as a whole to a local contractor through tender. It was built by Nuri Demirağ and Partners, called Simeryol.[27]

The reinforced concrete bridges of this line had been subcontracted to SAFERHA.[28] According to Akkaya, this railway project was a "School for Engineers". Feyzi Akkaya[29] and Sezai Türkeş were working in SAFERHA. The bridge designs were carried out by Akkaya, and site works were managed by Türkeş during 1937–1939. The bridge at 441 km (ACH-25) was a deck arch with a span of 45 m. During its construction, the scaffolding arch was distorted, and Akkaya was called to the site to solve the problem. Another bridge was a half-through arch bridge for vehicular traffic at 386 km (ACH-34), designed for road crossing (Figure 3.11).

In addition, two-deck arch bridges were constructed at 374 km and Sansa (ACH-22). Bridge at 374 km is a two-span railway bridge with 50 and 35 m spans. Bridge at Sansa is a three-span bridge, and it is the longest span of 35 m and a width of 8 m (Figure 3.12).

Akkaya also provides information on the Diyarbakır-Cizre line bridges which had been designed to have a metal superstructure; however, due to war, steel could not be provided by Germany. Therefore, these two bridges were finished as concrete

Figure 3.9 51-m span Soğanlı bowstring bridge of Irmak-Filyos line (author).

Figure 3.10 One of the deck arch bridges of the Irmak-Filyos line (author).

bowstring bridges. The span lengths were halved by constructing an additional pier in between the existing piers.

Railway steel bridges were mostly truss type with few examples of underspan bridges. The lattice type, for example, has not been used at all in bridge structures,

Figure 3.11 389-km Sansa bridge dated 1938 during the mourning of Atatürk. A black flag has replaced the national flag (STFA archive).

Figure 3.12 Sansa (389 km) bridge (STFA archive).

even it was used as a non-structural system in vernacular architecture, like the window closings of mosques and walls of authentic aerating structures, called Serender. Some examples of lattice truss use are the timber bridge in Jordan crossing over the Jordan

Figure 3.13 Palu bridge crossing over Murat river with temporary repair (İBB Atatürk Library).

river[30] and the lattice used for the repaired bridge of the railway line whilst crossing over the Euphrates in Palu crossing (Figure 3.13).

Another different truss type used was the Howe system, in which braces were timber and vertical ties were metal bars. This system was described in the 1940-dated railway construction book by Meissner.[31] One of them was used in the Hedjaz railway bridge, and the other type was used for a road bridge in İzmir province.[32]

3.4   REPUBLICAN ROAD BUILDING

Nafia classified roads into two groups: state and province roads. State roads were the main arterials, and their width was usually 5 m for vehicles and 1-m sidewalk on either side; however, this section was modified depending on local conditions and existing roads.

Nafia revised the procedure for constructing and maintaining the provincial roads in the early years of the Republic as identified in three periods by Nafia (1933). In the first period from 1923 to 1926, state roads were financed by the treasury and directly managed by Nafia. Province roads, which belonged to the local districts, owned and maintained by Provinces. The finance was provided by the "road tax system" in which locals paid cash for or supply labour for road works.

In the next 2-year period from 1927 to 1929, the Law on the Unification of State and Province Roads[33] was in force. All roads were constructed and maintained by the provincial administrations. Engineers and technical officers of provinces were appointed by Nafia. Finance was provided by the treasury and the road tax system, the total amount distributed to provinces as determined by Nafia.

An important downside of this system was that the provinces did not take responsibility to maintain the state roads and the income was diverted to different expenditures. Therefore, a new 1929-dated Roads and Bridges Law (no. 1525) brought changes to the tax system starting the third period.

In 1929, the practice of unification of state and provincial roads was abandoned, the roads designated as "national roads" were to be constructed by Nafia and province roads constructed by provinces. Law number 1525 also regulated the use of roads' tax and discounted the total tax amount, but decreased the province fraction and prevented the tax to be used for another expenses.

A noteworthy improvement for road building was the establishment of the Ford automotive factory in İstanbul in 1928. Ford could produce 50 cars in a day. However, this factory had to close in a short time as a result of WWII and the economic crisis.

Finance and Management of the Works: The finance for road building was mainly provided through the road tax system. The 1923 programme also proposed another income method in which road users or vehicle owners would pay a depreciation value for the roads, but this was not implemented (Figure 3.14).

Figure 3.14 **Preparation of lumber with hand frame saw for bridge construction near İzmir, in 1937 (1974 KGM Calendar album).**

Road building and especially the tax system were regulated through the laws which were revised frequently to adapt to the changing environment socially and economically. The first law, dated 1921 Road Tax Transfer, was followed by 1925 Road Tax Law, 1927 the Unification of Roads Law and 1929 the Law of Roads and Bridges.[34]

The 1921-dated law required all men aged between 18 and 60 to be eligible for the tax. Those with special needs with a medical certificate, people in poverty and those under military service were exempted from it. The tax was collected by the local head (Muhtar) and their councils. The tax was collected in cash, and those who could not pay were asked to labour on road repairs and constructions within their district.

In 1925, the road tax law was revised to prioritize the work physically in the projects rather than the cash payment.

The 1929 law mainly defined the borders and responsibilities of Nafia and provinces and made some revisions to the road tax system. The road tax was collected in labour or cash from those who did not want to work in person.

The yearly tax was equivalent of 10 days of wage of a normal worker.

The road tax system was vital for road works as a source of income or manpower until it ceased in 1948. The system had many impacts on social life and even political platforms. For example, there were always discussions about its implementation and its fairness. Long debates took place in public on whether including women in the tax system was required to contribute, as they were considered equal in whole rights. The tax also encouraged some families to have five children so that they could be exempt from it. In the 1945 elections, the road tax system was a hot topic for the opposition party and was even said to be regarded as a main factor for them to win the elections.

Road Projects: The first Republican roads were constructed by Provinces, and many were short strategic roads like Diyarbakır-Siirt road, Ankara-Environs and İzmir-Hinterland connections. Nafia's road projects started in 1929 with the first Balıkesir-Balya-Çanakkale, Hopa-Borçka and Trabzon-İran transit roads.

Balya-Çanakkale road was 160 km connecting Çanakkale to inner lands and Balya and further extended to the already existing Balya-Balıkesir and Edremit roads on the coast.

Balya was an important mining centre;[35] its vicinity was relatively straightforward and accessible for road construction and was strategic location with its boundaries to Aegean and Marmara Seas and the Dardanelles Strait. The leaders of the Republic had defended their land here only some few years earlier (Figure 3.15).

On its completion, the road had been inspected by the Chief of Warfare very thoroughly.[36]

This road was given heritage status in 2018 for its historical value as "the first asphalt road opened on tenth anniversary of the Republic, 29 October 1933".[37] Three of its bridges, Güngörmez, Seklik (Azapdere) and Müstecap, are also heritage registered. There are in total seven bridges on this road; Big Agonya, Small Agonya, Çandere and Nişankaya are the remaining.[38]

The next road was connecting the coastal city, Hopa to Borçka district, which was 30 km towards the south inland intending to connect eastern border cities to the Black Sea Region. It was close to the then Russian border, therefore built for

Figure 3.15 **Cropped from a detailed map of the Turkey showing natural recourses like mines, forestry as well as transportation infrastructures and provinces prepared by Halit Ziya Türkkan in 12 February 1931.**

strategic as well as economic reasons. Borçka (MTL-04) bridge is the only bridge on this road.

Trabzon-Gümüşhane-Erzurum-Karaköse road, also called Trabzon-İran transit road, 610 km road connecting Black Sea to Tebriz in İran opened in 1936. Part of the road was pre-existing in poor condition and was reconstructed to increase the Trabzon harbour facilities and provide a connection and good relations with İran. İranian Shah Pahlavi had travelled along the route during his visit in 1932. The Horasan (BWS-20) bridge completed in 1940 is on this road.

This road had very harsh conditions and demanded high maintenance. The government had bus services on this transit road and had to keep the road open all seasons. This was an initiating factor for the concept of all-season accessible roads and their maintenance in Turkey (Figure 3.16).

Ankara-İstanbul road was surveyed in 1931, and the layout was determined. The road was connected to Beykoz in İstanbul.

1923-dated Nafia programme prescribed roads in macadam technique.[39] The first major asphalt road, Çanakkale-Balya, was finished in 1931. The streets of major cities, like the new capitol Ankara and İstanbul, met with asphalt in earlier years.

The width of the bridges was initially designed to provide a comfortable passage for two cars with a 4.8-m-wide road and additional 60-cm sidewalks, a total of 6 m. The decks were waterproofed, and a 15-cm asphalt layer was used as the road surface.

After budget revisions, the road width was reduced to 3–4 m, and the sidewalks were cancelled in some bridges.

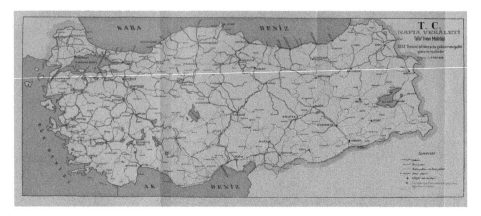

Figure 3.16 The road network in 1933 was given as an appendix to Nafia (1933).

3.5 REPUBLICAN BRIDGE-BUILDING

Besides the obvious need for bridges to cross obstacles, they very often were of strategic and political importance.

Garzan (BRD-15) was the first bridge constructed by the Republic, connecting Diyarbakır to Siirt and further extending to the Middle East. At that time, the borders were still not determined, and Mardin, now part of Turkey, was outside the border during the Lausanne negotiations.

There were many discussions during the first sessions of the Republic's parliament about this bridge. Parliamentarians from the region stated: "Thousands of animals, sheep, donkeys and people have lost their lives here. Eighty governors came and could not manage to construct the bridge. We claim to rule this place. How are we proving our reign?"[40]

This masonry bridge was constructed under difficult conditions and had to be built twice since the first bridge was damaged when its scaffolding was destroyed during flood. However, this dauntless generation turned this disaster into a useful lesson. It is important to note the mindset, that this failure was not hidden, rather published in their magazine openly and treated as "lessons learnt" (Figure 3.17).

Adagide (BWS-01) was the first reinforced concrete bridge of the Republic and opened in 1925. It connected the town Adagide to Ödemiş, where there was a railway station. Adagide became a tobacco warehouse, and even the base prices of tobacco were declared there. Tobacco agriculture was extremely important for the development of the region. Especially the tobacco monopoly[41] ruled and managed by Regie was ended by the new government at this time. The Adagide bridge was a bowstring type designed by Mehmet Galip Sınap who was a teacher between 1919 and 1922 at the Engineering School in İstanbul (Figure 3.18).

Designs: The designers of the bridges are not named in Nafia publications; however, some individual cases are known from other documents.

Figure 3.17 Garzan bridge, the first bridge of the Republican era from Nafia album (İBB Atatürk Library).

Figure 3.18 Adagide bridge, the first reinforced concrete bridge of the Republican era, from Nafia album (İBB Atatürk Library).

Bridge designs were carried out by contractors until 1929. For example, Irva (ACH-01) and Kirazlık (ACH-02) used the same design which was prepared by a foreign company.

Regie had prepared designs for some projects, and Nafia might have used these existing designs (Figure 3.19).

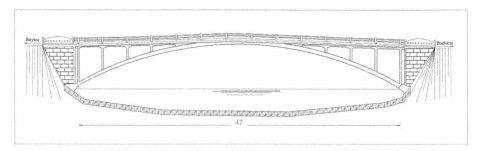

Figure 3.19 Irva bridge sketch (author).

Figure 3.20 Kirazlık bridge from "Turkey in Pictures" album dated 1937 photograph by Othmar Pferschy (TBMM Library).

Nafia improved on the pre-existing designs, for example, Kirazlık bridge was based on the previous Irva bridge design; however, the parapets were modified with concrete balustrades instead of metal railings and the foundations revised to concrete piles rather than timber piles. Nafia was also keen to use resources effectively, such as by increasing the span lengths with the use of cantilevers over the piers, in order to gain more in bridge total length. For example, Çarşamba bridge, a beam-type cantilever bridge, had its last spans cantilevered over piers, thus achieving 7 m more on both ends.

The design approach to structures was holistic with structural systems chosen according to ground conditions. Nafia was very cautious with foundations, adapting designs for the capacity of foundation elements. Structural systems were revised when ground conditions were found to be under capacity. For example, Silahtarağa bridge's abutments are designed with a hollow section in order to avoid lateral loadings to piles. The embankment designs and river bank treatments were also taken into consideration, during the design of the bridge.

Nafia had an open-minded attitude, it was important for them to be up to date and follow the latest technological improvements, as emphasized regularly in their publications. Pretensioning of reinforcements and Freyssinet technique for recompressing the arch bridges are some of the techniques applied as an early application of these methods internationally (Figure 3.20).

Experimental and innovative designs were also observed, like in an early structure, Güzelhisar bridge, dated 1927. Design was a concrete replica of screw pile for piers, which was used in many metal bridges. This design was planned to be used as a standard as explained by Nafia; however, no further application was observed.

1928 Akçay, 1928 Aslan, 1931 Müstecap, 1932 Nişankaya, 1933 Fevzipaşa and 1933 Aksu were sophisticated designs of their types. For example, Aksu bridge was designed with a hybrid system of a bowstring main span which continued as cantilevered beams over the piers to the adjacent spans.

After 1929, design works were mostly carried out in house by Nafia, and construction was the responsibility of the contractor including the design of auxiliary structures. For example, 1933-dated Aksu bridge design details and the calculations were published in the report and described as "beautiful technique and artefact designed and made by Turkish Engineers" (Figure 3.21).[42]

The diverse bridge types had been constructed during 1923–1940 with 20% bowstring, 30% arch type and 10% metal bridges with the remaining beam and unique bridges. Even though some bridge types are rare, like movable bridges, mainly due to the topographic conditions, Nafia was open to new types like the Müstecap frame bridge and Kemah suspension bridge. The selection of types changed over the years: the use of bowstring arch was replaced with beams as their spans increased, especially with prestressing. The metal-type bridges became more common over time (Figure 3.22).

In 1932, the Kömürhan arch bridge with a 109.6 m span was tendered to contractor including the design of the bridge. The preliminary studies of this bridge

Figure 3.21 Aksu bridge dated 1933 (Photograph: Bob Cortright, Bridge Ink).

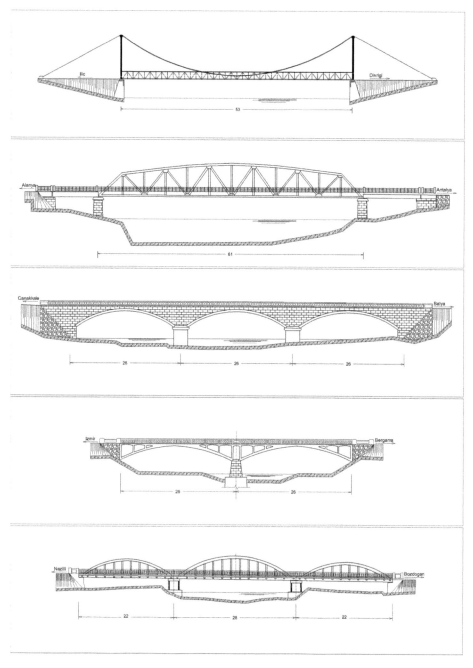

Figure 3.22 Examples of the diverse types of bridges constructed during the early Republican era (author). From Top to Bottom: BWS-04.Menderes Bridge, BRD-15.Garzan Bridge, ACH-07.Bakırçay Bridge, ACH-05. Büyük Agonya Bridge, MTL-02.Manavgat Bridge, MTL-05.Kemah Bridge.

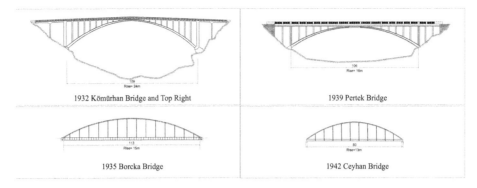

Figure 3.23 **Longest bridge spans constructed during the early Republican era (author). Top Left: 1932 Kömürhan Bridge and Top Right: 1939 Pertek Bridge Bottom Left: 1935 Borcka Bridge and Bottom Right: 1942 Ceyhan Bridge**

were carried out by Halit Köprücü from SAFERHA. However, the span range was beyond the local expertise, and this was probably the reason the bridge had been contracted out. Then, the same procedure is repeated for 1939-dated Pertek bridge, an arch with 106.9 m span, which was contracted to local company Aral, that had the design made by Emile Mörsch.[43] In between 1932 and 1939, remarkable bridge projects were achieved by local resources with promising designs, for example, a through arch of 72-m Körkün (ACH-12) was constructed with timber truss scaffolding, Pasur (ACH-13) deck arch bridge was constructed in a remote location with a 50-m span, and with a 62-m arch span, Keban Madeni (ACH-19) bridge was constructed with half of the scaffolding built vertical and rotated to form the arch (Figure 3.23).

The longest beam bridge was 1927-dated Kırmastı (BRD-02) bridge, and soon in 1931, it was overtaken by Çarşamba (BRD-03) bridge. Kırmastı bridge was a continuous beam type of 120 m total length, and Çarşamba bridge was cantilever beam type of 274 m total length.

Architectural Design: Public works had a mission of "the political and ideological charge of architecture in the service of nation-building" as stated by Bozdoğan.[44]

In terms of bridges, the engineering and aesthetic design parameters are naturally separate from each other in contributing to and affecting the edifice. Engineering design of the bridge is mainly determined by the site or project requirements and technology, and the design can consist of only the basic geometry without any architectural feature. On the other hand, if a bridge has an importance beyond its basic transporting function, like a landmark or icon bridge, then the architectural design may determine the appearance and the structural design will also serve this purpose. Therefore, the extent of aesthetic input can be from 0% to 100%, depending on the sociopolitical factors and the decision of the authority.

In regards to early Republican bridges, they certainly were expected to represent the ideas and principles of the new Republic, such as embracing modernity or demonstrating an evidence of the rich cultural background. In some cases, bridges have a purpose beyond their engineering function such as representing power and authority.

For any bridge, a good appearance may come by default with the design as a result of a structurally honest system and good proportioning which leads to a slender and orderly structure together with careful detailing. This approach was consistently observed in Nafia documents, in describing the bridges.

Architecturally, the first years of the Republican era were a transition period, called Ottoman revival. This was followed by a short period of new modernity after 1930. The later period was the engagement of vernacular and historical references in design from 1939 to 1950. However, as Bozdoğan states, even when styles were determined, indeed "the deep and sustained ruling motive" behind all was the building of a new nation.

Bridges that were purposely designed with architectural consideration will be mentioned in this chapter. The individual bridge descriptions are presented in related chapters of this book.

1925 Etlik and Ziraat bridges in Ankara are the representative of the first period,[45] that is transition period, designed by architect Kemalettin,[46] who was engaged by the Ankara government.[47] These two bridges used the same design except different widths (Figure 3.24).

Bridge is pleasing with the arch spans, repeating patterns and ornaments along the superstructure. The arch is five-centred, and the middle portion, nearly half of the span, is flat and resembles the basket-handled arch, which is a three-centred arch having a crown with a longer radius. The arch is purely an architectural feature whilst the main structural element is a reinforced concrete beam. Ziraat bridge has been demolished; therefore, the still-standing Etlik bridge remains as the only edifice for neo-classic architecture in bridges in Turkey.

In the Nafia report (1933), where bridges are described, there is frequent reference to the aesthetic appearance. There was a mature mindset that the bridge aesthetics were not alone left to architectural design or as decoration on top of the bridge design. The parapets of the bridges were always given good attention. More obvious evidence for this awareness can be seen in a letter[48] written to Nafia by a contractor, who was applying for proficiency in İzmir-Bergama road over

Figure 3.24 Etlik bridge of the Republican era from Nafia album (İBB Atatürk Library).

Gediz bridge. This road was treated as a touristic attraction road, and the specific aesthetics concern might have been for this reason. In this 1930-dated letter,[49] contractor included the architect as Semih Rüstem[50] with his references listed as Bakırköy Basmane bridge, Kadıköy Yoğurtçu bridge and Kağıthane Sünnet and Silahtarağa bridges in İstanbul and Aslan bridge on Ereğli-Devrek road.[51]

Amongst them, Basmane was a unique bridge of bowstring type with a trussed arch.[52]

1925-dated Yoğurtçu bridge, which is known to be designed by Semih Rüstem, is a concrete superstructure designed to be continuous for the length of the bridge with the parapets, posts and bollards featuring architecturally. 1928-dated Küçüksu and İstinye bridges are similar with the same approach of the simple structural system and heavy architectural elements other than the main structural parts. These are modern bridges in overall design with structure and architecture blended to create balance, harmony and even contrast, all to serve the overall appearance of the bridge.

1929-dated Ankara (BWS-03) bridge in Zonguldak and 1932-dated Silahtarağa and Sünnet bridges in İstanbul are all bowstring bridges with architectural features. Bowstring design itself was considered aesthetically pleasing by Nafia, and these three amongst them are also outstanding with their architectural features since they are also suited to the urban landscape. They have lighting poles atop of adorned pedestals.

1932-dated Göksu (ACH-14) bridge in Beykoz, İstanbul, was a notable bridge in İstanbul, on a main arterial road crossing the Göksu river with a basket handle arch and unique parapets. The arch of the bridge had a flat beam section in the middle resting on cantilever arch extensions.

Aslan bridge on Zonguldak-Devrek road has a remarkably slender deck arch with two ribs and unique parapets ornamented with star-shaped openings. Akçay (ACH-03) bridge in Aydın is another bridge with ribs and is also slender and appealing.

It is noticeable that all these bridges except Aslan and Akçay bridges were constructed by municipalities or provinces, that is local authorities. The reason might be the budget allowance or bureaucracy of the local authorities being more flexible to engage architect in design or architects could be already employed in the organizations. At the same time, the bridges being located at city centres or in their vicinity probably resulting in giving them additional architectural attention.

Parapet design is also a very helpful guide in understanding the conditions and technology in building the bridges as well as the ownership and attention given by the relative authority. The parapet is an element required for a bridge to function for the safe crossing of pedestrians and vehicles and to withstand vehicle collision to some degree. Therefore, they are good indicators of some design aspects, like importance given to a bridge, resources and appearance (Figure 3.25).

All the bowstring, metal and most of the beam bridges had metal parapets. A common example can be seen in Akçaağıl (BWS-9) bridge made of three rows of simple iron rods. There are also some exceptional designs like those of Manavgat (MTL-2) and Yalakdere (BWS-14).

Standards and Design: The design codes for bridges were not available until 1929. Projects were designed as per French or German codes as a choice of the contractor and/or client.

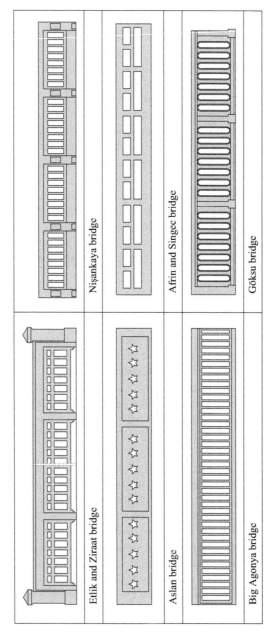

Figure 3.25 Concrete parapet details from various bridges built during the early Republican era (author).

On 2 June 1929, the law for "Roads and Bridge" construction was introduced, and since then, bridge-building was managed by a specific department constituted to prepare bridge projects.

German standards were applied for the bridge designs and exemptions or revisions accepted when needed for the local conditions.

The German codes were translated to Turkish and used to specify loads, calculations, strength checks and design live load for bridges. For example, for the live load in design, a load with 7.5 tons in the front axle and 5 tons at each rear wheel had been applied with a total load of 17.5 tons.

The earliest publication about bridges in the Ottoman language was the book written by Andre, chief engineer of "Ponts et Chaussées", as an appendix to the book on engineering by Leclercq. The translation was made by Ahmet Şükrü (Figure 3.26).[53]

The known first publication dedicated solely to bridges in Turkey was by Mehmet Hulusi, an engineer and academic, published in Ottoman language during the years 1893–1894. This book was in five volumes. The first three volumes were for masonry, timber and metal girder bridges, and the fourth and fifth volumes were on harbour construction and movable bridges. It is not known if these publications were translated partly or directly from foreign languages.

Others were Mehmet Fikri (Santur)[54] books in the Ottoman language for iron bridges in 1920, timber bridges in 1925 and suspension bridges in 1937. The book "Reinforced Concrete Bridges and Design" was published by Ahmet Ihsan in 1931.[55]

Later, Ali Fuat Berkman translated 1930-dated German codes and specifications for iron, timber and railroad bridges. These three books were published by İTÜ in 1950 as a second publication, but their first publication date is not known to the author. Berkman also translated the book on the arch bridges by Strassner.[56]

Books, mostly as translations, continued to be published, for example, Peynircioğlu[57] published three books specific to bridges in 1941.

Another noteworthy book was the translation from Meissner,[58] who was chief engineer of the Hejaz Railway. This 1940-dated book was written for local conditions for railway line construction and is about the substructure and investigations.[59]

The earliest Turkish standard[60] specifically prepared by Nafia for bridge-building found by the author is named "Technical specifications for the Road and Bridges, base levelling and road and masonry construction to be built within the Republic of Turkey" dated 1933 and published in İstanbul. This standard covered all stages of bridge construction and specifically prepared for local conditions. It has four sections. Chapter 1 describes the construction materials with their properties and preparations. Materials covered were stones, gravel, cement, gypsum, lime, brick, asphalt, bitumen and timber. Section 3.2 describes the application of the project on-site, Chapter 3 is for the road and pavement design, and the last Chapter 4 is the construction manual for timber piles, sheet piling and cast iron drainage pipes.

The standard focuses on stone masonry construction, and also there is basic information for plain concrete construction. Nearly all elements of the bridges are included except bearings, joints and steel at all.

A later edition of the technical standard in 1938 named "Technical Specifications for the Construction of Large Bridges"[61] and published by Nafia. The specification

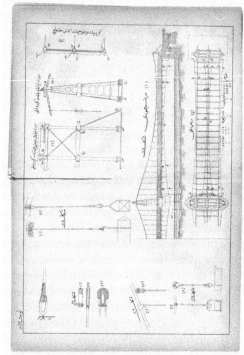

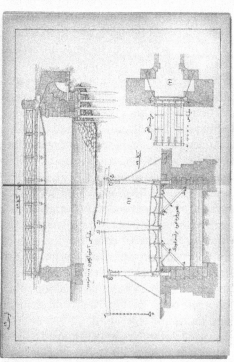

Figure 3.26 Pages from Ottoman bridge Building book written by chief engineer Andre (Library of Congress, Sultan Abdul-Hamid II Collection. 1842–1918).

Figure 3.27 **Marble Name Plaque of the Karabekir and Sakarya bridges (author).**

covers all elements of a bridge excluding the same as the previous version, however including reinforcing steel. In these standards, a marble plaque was required to be placed on the bridge with the name and date of the bridge. The slab should be made in white- and the inscriptions in black-fixed paint (Figure 3.27).

Substructure Works: Bridge-building is often the representation of the highest technology and standards in the country in terms of the construction industry. Bridges as a structure are divided into two parts both for design and also for construction. The superstructure is the part of the bridge that spans between the supports and provides the running surface for vehicles and pedestrians, and substructure is the remaining parts, namely bearings, pier, abutments and foundations. The superstructure is defined whilst describing the bridges in the related chapters. The substructure works will be summarized here.

Nafia usually mentions the ground conditions regarding its type and capacity whilst narrating the bridges; however, it is often not clear how this was determined. Bakırçay, Silahtarağa, Aksu and Sakarya bridges[62] are known to have boreholes for investigation before their construction, and in a few cases, where the site was not reachable, boreholes were carried out during construction. The ground conditions were taken into consideration for every bridge before concept design and bridge types were chosen accordingly. In case the ground investigation was not complete, the bridge type was chosen with that taken into account. For example, cantilever-type bridge, also called Gerber girders after the German engineer[63] who invented and patented them, would compensate for differential ground settlements were chosen for weak soils or when in doubt, whereas, in the Müstecap (BRD-18) bridge, since the ground was known to have enough capacity, continuous superstructure was chosen for the whole bridge length. Another case was found in a letter[64] in the archive as written by an authority, for the price increase because the abutment on the east side had to be piled, as the ground conditions were found to be worse than the tender project for the İsaköy (ACH-21) bridge. Akkaya provides useful information regarding groundwork.

He invited his university teacher, Fikri Santur, to their design office whilst he was working in the SAFERHA office in İstanbul. He showed a design to his teacher and further provided comments on the groundwork[65]:

all the calculations from top to bottom are done with 3 per cent error. When we come to the foundation, which we will rely on to carry this structure, we don't have any calculations... An old engineer walks on the soil pressing his heels and appraising "this soil can carry 2.5 kg"...

In saying this, Akkaya offends his teacher badly who exits the office leaving his coffee behind. Then, Akkaya adds:

... At that time [1933], neither Terzaghi's[66] name was heard of, nor the concept of "Soil Mechanics"... decisions were normally made by the use of half experience and half instinct, by wondering, and by hammering iron on the foundations and checking them ...

...years later Terzaghi became famous. Soil mechanics science improved as an engineering branch and foundation design based on calculations.[67]

For excavations, a common method was the use of a sheet piling inserted into the ground. Sheet piling was the plate piles, driven side by side and locked to each other with interlocking edges preventing water flow. Then, excavation starts inside the resultant manmade box. As a Nafia (1933) report explains:

The Majority of rivers in Turkey, except the ones like the Fırat (Euphrates), have a very low level of water flow. Commonly, the riverbed consists of sand and gravel, this results in water flow accumulating in the excavated foundation pit, making excavation very challenging. The worldwide solution for the foundation in deep water is the pneumatic caisson. However, pneumatic caissons were not in practical use in the country...

The pneumatic caissons are only rarely needed for bridges. They are not feasible and it is expensive to have this technique in the country. In very challenging rivers where the water is deep, the excavation is carried out by sheet piling...

In terms of piles, most of the early Republic bridges have been constructed with driven concrete piles. The earliest bridges with concrete piles were the 1926 Etlik and Ziraat bridges in Ankara. The first pile-driving machine was brought to Turkey by SAFERHA, as we learn from Akkaya.[68]

Timber piles were utilized for only five bridges out of 48 bridges introduced in Nafia (1933) report.[69] These bridges were 1926 Irva, 1932 Koyunağılı, 1932 Silahtarağa, 1932 Sünnet, 1932 Manavgat and 1932 Vezirhan bridges.[70] Notably, with the exception of Irva which was probably designed by private company, timber piles were not used in bridge-building until 1932.

Nafia report (1933) specifies timber piles to be used in submerged foundations as an improvement in bridge-building in the country. This report praises timber piles and describes them as used "since ancient times and proven safe without any problems". The preferred types were oak and juniper; however, these were often not available in the desired dimensions. Amongst other timber types, pitch pine in particular was also proven to be durable in water.

Although this report also states that "Driving reinforced concrete piles in foundations is also beneficial in some cases", many disadvantages in comparison to timber pile option were listed. One of them was the concrete pile being three times heavier than timber and requiring more powered machines. In addition, it was a considerable challenge to transport and assemble the pile-driving machine. Another was the time needed for concrete piles to be cured and harden, whereas timber piles were ready to use.

Nafia's approach was utilitarian in engaging timber piles as it was cheap and practical to utilize as an option whenever its use was feasible. Especially when the foundation was already excavated below the water level, the timber piles were a suitable solution.

Material: Nafia declared timber bridges as not suitable for the needs of the public and their policy of "building for longevity". They also had a preference to build bridges of masonry and its derivatives like concrete and reinforced concrete, describing it as "new material", which required less maintenance and was more durable compared to metal bridges. Additionally, it was available in the local market and could be constructed with local labour.

Even though Nafia defined the favourite material as concrete and commented on the preferences, the options were not many and the material used for bridge-building was based on the availability. Many bridges were still built of stone and timber, and more rarely steel where there was an unavoidable need.

Sand and gravel could be obtained from the rivers and creeks nearby the bridges, and cement was produced by local factories.

Reinforcing bars were still imported; however, "a number of reinforcing bars weren't excessive to consider", as explained by Nafia.[71] The term "armatür" was used directly from French for the reinforcement of concrete.

The favour for concrete is understandable as the worldwide propaganda of concrete as "everlasting material" had already started. Some disadvantages of concrete bridges were the longer duration of construction and higher dead load leading to greater dimensions and larger foundations.

As Aslanoğlu[72] explains, despite the urgent need for reconstruction of the country devastated by war, the building industry of the early 1920s consisted of only three sawmills, two brick factories and two cement factories.

The cement industry in the early Republican era consisted of one company, which formed by the merging of Arslan and Eskihisar factories in 1920.[73] The increase in demand soon caused shortages in building materials. Especially, the ingredients of concrete, the cement and steel rods, had to be immediately obtained to sustain the renovation.

In the year 1926, a cement factory was established by the municipality of Ankara. This was followed by the establishment of the Kurt cement factory, Yunus cement factory and Zeytinburnu cement factory in 1929 by private enterprise.

1932-dated Kömürhan was also the first bridge in Turkey to be constructed with the imported high-strength concrete also known as "super cement". Later, the Kartal factory started production of "super cement", and their product was first used in Manavgat (MTL-02) bridge.[74]

Stone and brick masonry bridges had been used since ancient times. One of the reasons they were abandoned in time and replaced with concrete was their requirement for a huge volume of material for spandrel infill, and another was the limited span length. However, masonry bridge-building continued for some years declining as the construction industry replaced masonry with concrete.

Bridge-building in the early Republican era was also a continuation of cumulative expertise, especially for local practices. Therefore, timber and stone bridges continued to be built. Stone bridge practice was abandoned with time as it could not cope with the conditions and demands of the industry. Quarrying, stone cutting and transporting were all very labour- and time-intensive compared to concrete. Timber bridges were easier to build by local people with accessible material, and they were used for some years until natural timber processing became limited due to scarcity of resources, needed special expertise and also became expensive compared to the industry.

Steel was expensive in its initial price and needed regular maintenance. Therefore, it was more often used in railway lines where the site was remote and when it became a feasible option. Steel had to be imported from abroad, and its availability became an even greater problem during WWII.

3.6 BRIDGE NAMES AND OPENING CEREMONIES

Bridge names were chosen on a practical basis; usually, the name of the river or the closest landmark in the vicinity was used in the names. For example, Orman Çiftliği (BWS-02) bridge, also called Gazi bridge, named after the nearby farm, which is also called with both names. The few exception to naming after their location can be listed as follows: İsmetpaşa (Kömürhan), Fevzipaşa,[75] Karabekir,[76] Çetinkaya and Gazi[77] bridges. İsmetpaşa was chosen by Atatürk, Fevzipaşa and Çetinkaya bridges were named as a wish of the residents and Gazi bridge was so named by municipality.

Bridge openings were important formal events with public crowds. Long speeches were made by Nafia management and local officials, a triumphal arch erected at the entrance with the name of the bridge or a motto or even photographs of the minister/s, food might be delivered and a ribbon cut. In these ceremonies, modernism was emphasized, and women and children were contributing to the front of the display (Figure 3.28).

Kömürhan, Aksu, Gediz, Borçka, Singeç, Sakarya, Çetinkaya, Gülüşkür and Gazi bridges[78] are recorded to have been opened with ceremonies and the involvement of government authorities.

Railway openings were more splendid with a huge, people sacrificing animals, food supplied and long speeches given by Nafia and government authorities (Figure 3.29).

3.7 BOSPHORUS STRAIT CROSSINGS

The dream of a Bosphorus crossing was resumed after the Republication was formed. Nuri Demirağ came up with a proposal for a suspension bridge in 1931. He engaged the Bethlehem Steel Company which was involved in building the Golden Gate bridge. Despite his very determined efforts to promote this project such as dropping pamphlets by airplane over İstanbul, the authorities would not support him, and this proposal too was passed over. This bridge was proposed to be from Ahırkapı-Doğancılar shore, that is historical peninsula to Asian side, with 701-m span between towers (Figure 3.30).[79]

Then, the Krupp company prepared a proposal, engaging Paul Bonatz, who was at that time a professor at the İstanbul University, for another suspension bridge in

Figure 3.28 Opening of Aksu bridge (STFA archive).

Figure 3.29 Tenth year celebration of the Erzurum-Kars Railway line, with magnificent ceremonial arches for "victory and pride", as noted on the back of the photograph, with images of leaders and a model train at the crown of the middle arch (İBB Atatürk Library).

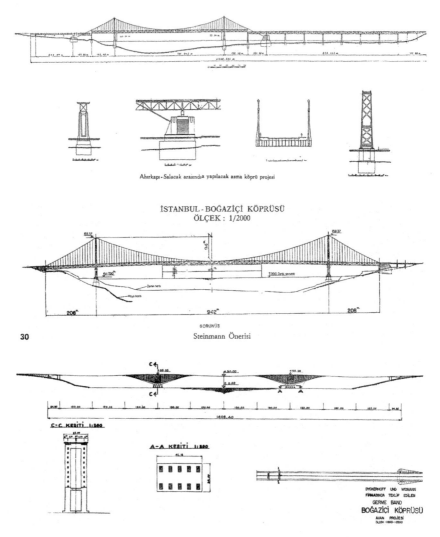

İSTANBUL - BOĞAZİÇİ KÖPRÜSÜ
ÖLÇEK : 1/2000

Steinmann Önerisi

Ahırkapı - Salacak arasında yapılacak asma köprü projesi

BOĞAZİÇİ KÖPRÜSÜ

Figure 3.30 (Top) Nuri Demirağ's suspension bridge proposal for the Bosphorus from İlter (1973). (Middle) David Steinman company as suspension bridge proposal for the Bosphorus from İlter (1973). (Bottom) Finsterwalder's Catenary-cantilever bridge proposal for the Bosphorus from İlter (1973).

1951 which was located in the place where the current bridge is today. In 1955, De Luew company was engaged for carrying out the preliminary works, and the project was put out to tender.

"Steinman, Boynton, Gronquist and London" firm managed by Steinman made proposals for a suspension system with a trussed deck and steel towers in 1959 (Figure 3.30).[80]

Another interesting and charming proposal was a design of Ulrich Finsterwalder as a Catenary-cantilever bridge by company Dyckerhof and Widman. However, his design was found to be prone to deflections and rejected. Even though the company revised and resubmitted the project, it was still rejected as it did not fit the landscape of the Bosphorus.[81] Then, the same company proposed a concrete suspension bridge; however, this proposal was not even considered as it was not abiding the tender regulations as explained by İlter (Figure 3.30).

The latest, although probably not the least, was a proposal for a torsional rigid deck by Max Herzog as a suspension system with a 1000-m span between towers for Bosphorus. This project might be the one submitted by company Dyckerhof and Widman. In the article, Herzog describes his project for a comparison of its cost with the trussed deck option of Steinman.[82]

Finally, the contract for the current bridge was secured with Freeman Fox Company for the design and consultancy. An international tender was then awarded to Hocthtief-Cleveland Joint Venture for the construction to start on 24 April 1970 formally.

Freeman-Fox had already designed many suspension bridges including the Severn and Humber bridges in Britain. Severn bridge, completed in 1966 with a main span of 590 m, was a forerunner model to the Bosphorus bridge. The Bosphorus bridge has a main span of 1074 m which was the longest span outside the USA. The original design had over 900-m span, but this was increased to construct the towers on dry land, thus reducing the cost and construction time of the bridge. It carries dual three-lane carriageways and created the first-ever road connection between the European and Asian parts of Turkey.

Freeman-Fox bridges, including the Bosphorus bridge, stand out with their streamlined box section, which was an innovative design to allow air to move around the deck rather than through it as in the case of truss decks. This aerodynamic shape leads to a reduced span to depth ratio of the bridges, thus producing slender and elegant structures.

Another new design feature was the inclined hangers. This has also been used for the Severn and later the Humber bridges. Each hanger is inclined rather than being perpendicular to the deck. This allows the hangers to carry some longitudinal force and further helps to increase the damping. Even though this intention worked, it led to another problem for hangers to be loaded and unloaded under live loading which resulted in fatigue problems. Designer William Brown has added some weights to the hangers to stabilize them.

Testing of this bridge could be considered as a rather dubious social experiment. Truck drivers were waiting in the long ferry queue when the announcement came of the opening of the bridge on the night of 15 October 1973, and they were let on to use the bridge. The trucks were then kept on the bridge for 1.5 hours for testing each loading condition.

Since its planning, the Bosphorus bridge has become part of public history, and its mark can be found everywhere ranging from identifying İstanbul in movies to Turkish money, tea and coffee trays, matchboxes, carpets and many more, but more importantly in people's lives and memories.

In Turkish culture, people believe that envy can cause harm or bring bad luck. This is called "nazar". Turkish people use "nazarlık", a blue-coloured, eye-shaped glass amulet usually hung at the entrance of the houses accompanied with the script "Maşallah" that means "with the will of God".

The Bosphorus bridge could not be left out from this treatment on its completion in 1973 when it became the hope, symbol and mark of the new Republic for its modern, prosperous and bright future.

A "Maşallah" for the bridge is located at its entrance.[83] This sculpture was financed by the Koç family according to their mother's will after Sadberk Koç expressed her wish saying that the bridge needs nazarlık.

3.8   PROMINENT BRIDGE ENGINEERS OF THE EARLY REPUBLIC

The main factor of any achievement is the people behind it; therefore, the engineers should be mentioned to complete the story of the bridges. Kemal Hayırlıoğlu, a chief engineer responsible for bridge-building in Nafia during 1928–1943, is introduced to provide a portfolio of an engineer from authority. A company called SAFERHA and three engineers, Halit Köprücü, Sezai Türkeş and Feyzi Akkaya, who were the main actors of the early Republican bridge engineering and were involved from the start of the timeline of the bridges covered in this book.

It is important to understand the cultural background behind their products and achievements for the recognition of the people. Eastern culture does not traditionally credit individuals, and accordingly, one should stay humble and not seen in the picture. For example, Sinan,[84] the sublime architect engineer, described himself as "mûr-i nâtüvân" meaning "poor ant" and refused to place his signature on mosques which he also believed belonged to God. Interestingly, however, he did place his signature on the inscription of Büyükçekmece bridge.

Even though the attitude to concentrate and value work, rather than person, may seem appropriate and expedient, its inevitable downside effects appear in the long term. When human factor neglected or not fairly recognized, the essential background and environment for the progress of profession does not form. Consequently, knowledge and experience accumulation does not take place which would lead to growth, as well phrased by, "standing on the shoulders of giants".[85]

On the other hand, the lack of this recognition for the responsibility and achievements of professionals appear as an over credit or blessing of the individual, who became a sacred figure rather than a leading role model.

The cultural and social environment during the Republic was different in this regard as expected, individuals were highly regarded especially for their works and their names were usually mentioned in the documents. However, the culture and attitudes change over time. There wasn't any specific biography for the architect and engineers except Sinan in the early Republican magazines and publications either. Furthermore, when a project was introduced, the engineer or architect involved in it was only named without any further information.

During the late Ottoman Empire period, the usual occupation for engineers was an engagement in the military due to unstable conditions or they went abroad for higher study, and the few who remained might have worked in the local industry.

The engineering industry, dominated by foreign investments, mostly seemed to operate as a separate private sector entity with few interactions with locals. The interaction between foreign engineers recruited as chief engineers or various positions in governmental offices, and the local engineers, as colleagues, was dependent on the organizational culture and environment. As stated by Martykánová[86], "It was not until the Turkish economy overcame the semi-colonial position it had fallen into during the second half of the 19th century, that local engineers could efficiently expand their field of action".

The engineers who had leading roles during the early Republican era were born in the Ottoman Empire and attended newly opened or revised schools for their education. The alphabet was changed in 1928 adopting Latin letters over Arabic. The metric system though was made mandatory with a special law from 1 March 1871, in all official transactions and for all subjects throughout the empire from 1 March 1874.[87]

The first engineers were graduates of the Engineering School, which was removed from military supervision and established as the Higher School of Engineers (Mühendis Mekteb-i Alisi) under Nafia in 1909. Later, in 1928, the school was transformed into the Higher Engineering School (Yüksek Mühendis Mektebi).

The engineers had common spirit, and they were hardworking, brave, committed and very determined, as they were empowered by the new, independent and victorious Republic. They were also attributed a higher rank by society, and this was followed by devotion to the profession which also brought recognition and esteem.

The devotion can be observed from surnames. The "Surname law" in 1934 required every citizen to adopt surnames, and engineers chose theirs accordingly. Their surnames give a good hint of their involvement; Köprücü (Bridgebuilder/Pontist), Demirağ (Ironnetwork), Akkaya (Whiterock), Demirci (Ironwork) and Betoncu (Concretermason).

Kemal Hayırlıoğlu (1891–1984) was a leading engineer of bridge activities as understood from the publications and memories of Akkaya. Hayırlıoğlu published papers[88] in technical magazines for the introduction of bridge works and presumably the author of the 1933 Nafia report which has been an important reference for this book. He was sent abroad to learn the latest methods and also to represent the young Turkish authority, for example to the International Exposition in Liege/Belgium in 1930,[89] where he presented Kömürhan bridge.

Hayırlıoğlu was a very bright student, graduated Engineering School with a high score and was sent to Europe for extra training by a special decision. After studying at the Technical University in Berlin, in 1914, upon the start of the First World War, he returned home to do his military service. In 1922, he resumed his education in Berlin. Although offered a good position in Germany, he returned to Turkey in 1924.

He worked in Ankara municipality as an expert engineer during 1925–1926, and managed the construction of facilities. Until 1928, he served as the Head of the Science Committee.

He was Nafia Chief Engineer for 15 years. On 1 August 1928, he was appointed as an Expert Engineer to the General Directorate of Roads of Nafia and a few months later was assigned to the administration of bridge construction. Until 1943, he served as the Chief Bridge Engineer and the Director of the Bridges Science Committee. In 1943, he was appointed as a member of the Expert Science Committee and continued working in the authority in a similar position until retired in 1955.

He also served as a teacher at Ankara Nafia Science School from 1928 to 1930 and was the President of the Schools Commission for Engineering School and the technical school for 10 years.

SAFERHA was an engineering firm that specialized in bridge and port constructions and built the most in number and significant bridges of the early Republican era.
SAFERHA was founded on 17 July 1928,[90] taking its name from the first syllables of
the three engineers who founded the company. They were pioneers in many aspects,
for example bringing the first pile-driving machine to Turkey.

Sadık Diri graduated from İzmir High School in 1912, then entered Engineering
School in the same year. Due to the Balkan and WWI, he postponed his education for
5 years and graduated in 1923.

Ferruh Atav graduated from İzmir High School in 1914 and finished Engineering
School in 1923. He then did internships in High School in Germany and various German companies and returned to Turkey in 1925. Later, he worked in the construction
of Çanakkale Ceramic company, of which he was also a shareholder.

Halit Köprücü graduated as an Engineer in 1923 and then worked for 2 months
for the İzmir-Aydın Railway Company, which was operated by the British at the time.
Following this, he started his career with his school friend Sadık Diri.[91]

He was a prominent bridge engineer, and some of the significant bridges known to
be his design are Körkün (ACH-12) and Pasur (ACH-13).

His innovative solutions were noteworthy to engineering problems like a jetty
structure for military use. The company was put in charge to construct the jetty with
limited resources including reinforcement and time. Köprücü built the jetty structure
in its final place on top of wooden piles above the water level. After concreting, the
slab was lowered in to the water by detonating the scaffolding and the structure lowered through guiding piles. The jetty structure was a waffle slab with an open bottom
and top, after its descent, concrete was poured inside the combs and a monolithic pier
was obtained. His design needed considerably less reinforcement for the structure, in
comparison to designs in which the jetty rests on concrete piles. He also patented this
innovation, and the design was applied to many jetty structures.[92]

Halit Köprücü later worked in KGM. He was interviewed for the Bosphorus crossing
in a 1952-dated Newspaper[93] in which he favoured an arch bridge over the tunnel crossing.

Ferruh Atav resigned from the company on 11 May 1937, then the company continued bridge-building works as SAHA until it went bankrupt in 1952.[94]

Akkaya mentions SAFERHA in his memoirs, and he declares that they, Türkeş
and himself, gained experience from positives, and learned lessons from negatives. He
then adds "we did not earn anything, but our spiritual gain was great. They seeded in
us the principles of unconditional quality, and finishing the work on time even under
difficult conditions even in the case of financial loss".[95]

The successor of SAFERHA in bridge legacy was STFA, which was founded by
Sezai Türkeş and Feyzi Akkaya, the company name was also derived from the first
letters of founder's names. These two eminent engineers have been leaders and role
models for engineers with their professional success, enthusiasm and stance. Their
contribution to the country and the engineering profession with many inventions and
first time applications in the field was recognized with the "Distinguished Service
Medal" for the first time in Turkey. Their motto implying the quality of the work was
"even if the concrete is to be buried under soil, it shall still look good".

Türkeş and Akkaya witnessed the Republican era starting from Nafia, SAFERHA
and then as STFA.

Figure 3.31 Sezai Türkeş & Feyzi Akkaya, Prominent Early Republican Engineers (STFA archive).

Sezai Türkeş was born in Lefkoşa, Cyprus, in 1908. He graduated from Engineering School in 1932. Then he worked at Nafia for 1.5 years in the port construction department. After 1.5 years of military service, Türkeş started working in SAFERHA in 1934 (Figure 3.31).

Feyzi Akkaya born in 1907, in Üsküdar, İstanbul, a grandchild of Ottoman statesman Tatar Osman Pasha, graduated from Engineering School in 1932, worked in the design department of Nafia for 1.5 years and went to military service, where he worked under Halit Köprücü on the Tekirdağ pier project. Then started working in SAFERHA technical office in 1935 (Figure 3.31).

Feyzi Akkaya has written his memories in a book *"Ömrümüzün Kilometre Taşları (Kilometer Stones of Our life)"* and also has written an 11 book set technical series, called "Şantiye El Kitabı (Pocket Books)" on numerous subjects. These books provide vital information and helpful advice to engineers. The style of the book is very friendly and engaging.

In an ironic "joke of fate", Akkaya said that Türkeş, Hamdi Hikmet Barkın and himself were the only students in their class who had to take a supplementary exam to pass the subject of "Concrete Bridges".

Later, when they were working for the Sivas-Erzurum Railway line, they designed and constructed some of the first Turkish reinforced concrete bridges. Their boss Tatar İzzet Bey, manager of SIMERYOL, would say "I did not have "simarname" (concrete) in my diploma" and made it clear that he had no sympathy for the concrete bridges.[96]

In 1935, Türkeş and Akkaya started their own partnership building many firsts in the country and were the successor of the foreign companies in many engineering projects like the ports of Kuşadası, Bartın and Ereğli, 154 kV Transmission line and the Kadıncık-II Hydroelectric Power Plant as well as the many significant bridges. During their time, Türkeş imported the first tunnel boring machine (TBM) to the country.

Akkaya[97] defines the situation of this taking over from a foreign company, who had even the machinery ready on-site for the project, when they were awarded the Ereğli Port project, as "selametlemek", meaning "to leave in peace".

NOTES

1. The term used for Public Works was "imar", means civilization and development excluding the social content. Term later changed to "Nafia", which can be translated as utility, to refer to public assembly at the governance level. Nafia also later changed to "Bayındırlık", means prosperous, in 1935.
2. İnan, A. (1989) İzmir İktisat Kongresi, 17 Şubat - 4 Mart 1923. In *Atatürk Kültür Dil ve Tarih Kurumu ve Türk Tarih Kurumu Yayınları*. TTK Press. Ankara.
3. Pamuk, Ş. (2012). *Ottoman Economic Legacy from the Nineteenth Century from: The Routledge Handbook of Modern Turkey*. Routledge. Oxon.
4. Pamuk (2012).
5. Feroz, A. (1993) *The Making of Modern Turkey. The Making of the Middle East Series*. Routledge London and New York. Page: 94.
6. İnan (1989) Page: 65.
7. Vekalet can be translated as 'power of attorney' and nezaret can be 'directorate'
8. Timur, T. (1971) *Türk Devrimi ve Sonrası 1919-1946*. Doğan Yayınları. Ankara. Pages: 75–123.
9. Nafia (1933) *On Senede Türkiye Nafiası 1923-1933*. T.C. Nafia Vekaleti Neşriyatı. İstanbul.
10. **Ali Çetinkaya** (1878–1949) was an Ottoman-born Turkish army officer and politician, who served eight terms in the Grand National Assembly of Turkey, including a period in 1939–1940 as his country's first Minister of Transport.
11. **Ali Fuat Cebesoy** (1882–1968) was a Turkish army officer and politician. In accordance with his will, he was buried to the backyard of a mosque near Geyve train station, where the first shots of the Turkish War of Independence were fired.
12. Tekeli, İ., and İlkin, S. (2004) used as a source and program was not available to author.
13. This program could not be found in the archives, its existence is accepted through the events, rules and decision taken by TBMM by Tekeli, İ., and İlkin, S. (2004).
14. A span (Turkish:karış) is the distance measured by a human hand, from the tip of the thumb to the tip of the little finger.
15. Feroz (1993) Page: 95.
16. **Nemlizade** is a local wealthy family. They involved mainly in tobacco trade in Samsun. They financed transportation projects like Samsun-Çarşamba railway and Değirmendere stone bridge in Trabzon.
    Zeki, M. (1930–1932) *Türkiye Teracimi Ahval Ansiklopedisi*. Cilt III. İstanbul. Page: 507.
17. Emin Sazak was a parliamentarian from Eskişehir and his sons established Yüksel Construction Company, one of the leading company in the construction industry in Turkey.
18. Atayman, M. S. (1967) *Bir İnşaat Mühendisinin Anıları, 1897-1918*. Baha Matbaası. İstanbul.
19. **Regie (Régie)** company was formed by a consortium of European banks and was the largest foreign investment and cooperation in the country. The capital of the company made up around 23% of total foreign direct investment in the Ottoman Empire in 1881–1914.
    Birdal, M. (2010). *The Political Economy of Ottoman Public Debt, Insolvency and European Financial Control in the Late Nineteenth Century*. Tauris Academic Studies. London.
20. Haykır, Y. (2011) Atatürk Dönemi Kara ve Demiryolu Çalışmaları. Doctorate Thesis submitted to Fırat University.
21. American Rear-Admiral Colby Chester who was sponsored by the New York Chamber of Commerce, tried to get concessions for railway constructions in Ottoman and Republican era.
22. Société Industrielle des Travaux (S.I.T) was a Belgian company.

23. The portions constructed by Nuri Demirağ has given differently in various resources. But he had constructed more than 1000 km railway line in total rail network. Nuri Demirağ's brother, Abdurrahman Naci Bey, was engineer and he had a company as well.

    Nuri Demirağ was an industrialist and entrepreneur of the early Republic. He was graduated from high school as a teacher, then started working in finance sector. With his first savings, he started business in cigarette paper called "Turkish Victory". Later, he continued his career as a contractor for construction works in the transportation network. In addition, he built factories like the Iron and Steel Plant in Karabük. Demirağ laid the first foundations of the aviation industry in 1936. He opened airplane workshop that evolved to a factory. The first domestic Turkish plane, built in İstanbul factories, flew to Divriği, the birthplace of Demirağ, in August 1941.

24. Babaoğlu, R. (2016). Türkiye Cumhuriyeti Nafia Vekaleti Devlet Demiryollari Samsun-Sivas Demiryolu Amasya İstasyonunun İşletmeye Küşadı 21 Teşrinisani Hatırası 1927. *Recent Period Turkish Studies*, 13:25–26, 183–234.

25. NOHAB (Nydqvist & Holm AB) was a manufacturing company based in Sweden.

26. Report (1937) Construction de Lignes de Chemins de Fer, Irmak-Filyos and Fevzipaşa-Diyarbakır. TBMM Library. Retrieved from: www.tbmm.gov.tr.

27. The name derived from the capitals of Sivas, Malatya, Erzurum and Yol, which means road. SİMERYOL was an important contractorship firm of the period. It was a community of contractors and its administrative council director was Tatar İzzet Bey. Ref: Akkaya (1989) Page: 77.

28. SAFERHA, a prominent company of bridge building.

29. Akkaya, F. (1989) *Ömrümüzün Kilometre Taşları: STFA'nın Hikayesi*. Bilimsel ve Teknik Yayınları Çeviri Vakfı. İstanbul. Page: 69.

30. İstanbul university archives date 1922 as Nehrü'l-Ürdün üzerine inşa olunan köprü. Retrieved from: http://nek.istanbul.edu.tr:4444/ekos/FOTOGRAF/90504---0049.jpg.

    The same bridge was also photographed by Gertrude Bell 1900, A129, Keeper of the Jordan Bridge. Gertrude Bell Archive Newcastle University.

31. Meissner, A.H. (1940) *Demiryollar*. Cilt1-Cilt 4. İTÜ. İstanbul. Translated to Turkish by Prof. Dr. Enver Berkmen.

32. Album (1929) Vilayet-i Hususi Idareleri Faaliyetlerinden. Hilal ve Cumhuriyet Matbaaları. İstanbul.

33. Tevhidi Turuk Law dated 22 June 1927 No: 1131

34. Turkish names for the laws: 1921 Tarik Bedeli Nakdisi, 1925 Yol Mükellefiyeti, 1927 Tevhid-i Turuk and 1929 Şose ve Köprüleri Kanunu.

35. Balya was known for mine resources since Roman times. It was mined by a French company during Ottoman reign and a rail line was constructed between Balya and Edremit.

36. BCA 030_10_00_00_155_90_7_1: Balya-Çanakkale yolu inşaatındaki yolsuzluk iddiaları.

37. T.C. Culture and Tourism Ministry, Balıkesir Regional committee for the Protection of Cultural Heritage. Env. No: 10.02/34-24.

38. Bridge codes are: Güngörmez (BRD-05), Seklik (Azapdere), Müstecap (BRD-18), Big Agonya (ACH-05), Small Agonya (ACH-06), Çandere (BRD-07), and Nişankaya (ACH-10).

39. **Macadam** is a type of road construction, pioneered by Scottish engineer John Loudon McAdam around 1820, in which single-sized crushed stone layers of small angular stones are placed in shallow lifts and compacted thoroughly. (https://en.wikipedia.org/wiki/Macadam).

40. TBMM Zabıt Ceridesi, Devre: 1, Cilt: 5, İçtima Senesi: 1, Doksan Beşinci içtima:1. XI. 1336. Cumartesi (1920).

41. **Regie** held the tobacco monopoly for 42 years, from 1883 until the contract was terminated in 1925. After the establishment of the Régie, all factory owners in the Ottoman Empire were forced to close their factories. On 26 February 1925, the Turkish parliament passed a law abolishing the Regie Company and officially nationalized the tobacco monopoly.

    Birdal, M. (2010) Pages: 15,151,165.

42. Nafia (1933) describe: "Aksu bridge is a beautiful work of science and art that is thought and made by Turkish engineers." The word "thought" used instead of "design" probably implies the concept design of the bridge.

43. **Emil Mörsch (1872–1950)** studied civil engineering at Stuttgart TH. He worked in the Ministerial Department for Highways & Waterways, and afterwards was employed in the bridge unit of State Railways. He joined the Wayss & Freytag company in early 1901. From 1916 onwards, Mörsch worked as professor of theory of structures, reinforced concrete construction and masonry arch bridges at Stuttgart TH.

   He published the first edition of his book 'Der Eisenbeton: seine Theorie und Anwendung', which underwent numerous reprints. This book set used as a standard in reinforced concrete for more than half a century. Book was well known in Turkey.

44. Bozdoğan, S. (2001) *Modernism and Nation Building: Turkish Architectural Culture in the Early Republic*. University of Washington Press. Seattle.

45. The First national architectural movement, also referred to in Turkey as the National Architectural Renaissance, or Turkish Neoclassical architecture. Inspired by Ottomanism, the movement sought to capture classical elements of Ottoman and Seljuk architecture and use them in the construction of modern buildings.

46. **Mimar Kemalettin** (Kemalettin the Architect) (1870–1927), was a renowned architect of the very late period of the Ottoman architecture and the early years of the newly established Republic. He was among the pioneers of the first national architectural movement.

47. İlter, İ. (1989) *Mimarlığımızın Neo-Klasik Dönemi ve Bu Anlayışın Köprülerdeki Tek Örneği: Etlik Köprüsü*. Karayolları Vakfı Dergisi. Ankara. Pages: 6–9.

48. Örmecioğlu, H. T. (2010) Technology, Engineering, and Modernity in Turkey: The Case of Road Bridges between 1850 and 1960. Ph.D. thesis submitted to Middle East Technical University (METU).

49. Petition of a contractor for document of certificate of proficiency, 1930. Source: Unclassified documents from the State National Archives-Republican Archives, KGM Fund, Binder no: 2020. Found in: Örmecioğlu, H. T. (2010) Technology, Engineering, and Modernity in Turkey: The Case of Road Bridges between 1850 and 1960. Ph.D. thesis submitted to Middle East Technical University (METU). Appendix. Figure H.3.

50. **Semih Rüstem** (1898–1946), is an Early Republican era architect who studied architecture at the Technical University of Budapest. In the early 1930's, he designed several buildings mostly under the influence of European Modernism but traces of these Hungary-related steps can also be followed in some of his designs.

   Gümüş, M. D. (2015) A Turkish Architect at Technical University of Budapest: Semih Rüstem. *Periodica Polytechnica Architecture*, 46:1, 38–45.

51. Bridge codes are: Basmane Bridge (BRD-17), Yoğurtçu Bridge (BRD-10) and Sünnet (BWS-07), and Silahtarağa (BWS-06) Bridges and Aslan bridge (ACH-04).

52. An impressive example of this type was constructed in 1927 over river Oued Mellèguein in Tunisia. The bowstring bridge on Mellègue Wadi in Tunisia was an engineering exploit designed by Henry Lossier for the Entreprises Fourré & Rhodes.

53. Günergün, F. (2005) Mektebi Harbiye'de okutulan mimarlık ve inşaat bilgisi dersleri için 1870'li yıllarda yazılmış üç kitap. In Mazlum, D. & Cephanecigil, G. (Eds.), *Afife Batur'a Armağan: Mimarlık ve sanat tarihi yazıları*. İstanbul. Pages: 151–163.

54. **Fikri Santur** (1878-1951) He graduated from Engineering School in 1899. In 1900, he started working in the same institution as Deputy Teacher. His main branch was Strength of materials and bridges. He retired in 1943 after 43 years at the same institution.

55. İstanbul Technical University Library Catalogue. Earliest publication dates for iron, timber and suspension bridges are 1920, 1925 and 1937.

56. Strassner, A. (1949) *Yeni Metodlar Cilt: 2 - Kemer ve Kemerli Köprü Statiği*. İTÜ Publications. İstanbul. Translated to Turkish by Berkman, A.F.

   Original book: Strassner, A. (1927) Neuere Methoden zur Statik der Rahmentragwerke und der elastischen Bogentrager. Aufl. Berlin: Ernst & Sohn.

57. Peynircioğlu, H. (1951) *Köprüler - Cilt* I. Teknik Okulu Yayınları. İstanbul. Sayı 5.

58. **Heinrich August Meissner** (1832–1940) Chief engineer of the construction of two lines, Hijaz and Anatolian-Baghdad lines from Germany. He spoke Turkish fluently. After completing the Hijaz line, he was awarded the title of pasha in 1904 by Abdülhamit ll. In 1918 Meissner moved to Germany but he returned back in 1924 as adviser on building and maintenance of railroads in new republic by the invitation of Atatürk.

59. Meissner, A.H. (1940) *Demiryollar, Cilt 1 – Cilt 4*. İTÜ. İstanbul. Translated to Turkish by Prof. Dr. Enver Berkmen.
60. Türkiye Cumhuriyeti Dahilinde İnşa Olunacak Şose ve Köprülerin Tesviyei Turabiyesine ve Şosa ve Kargir İnşaatına Dair Fenni Şartname (1933) Tecelli Matbaası. İstanbul. (No further information was provided about the publication).
61. Nafia (1938) *Betonarme Büyük Köprüler Hakkında Fenni Şartname*. T.C. Nafia Vekaleti Neşriyatı. Seri:1 Sayı:5
62. Bridges are: Bakırçay (ACH-07), Silahtarağa (BWS-06), Aksu (BWS-10) and Sakarya (BWS-15)
63. **Heinrich Gerber** (1832–1912) German civil engineer and inventor of the Gerber beam. Gerber beam is also known as cantilever beam, designed with several joints within the superstructure, therefore can compensate differential settlements between supports of the bridges.
64. BCA 030-18-01-02-82-26.6: Şile-Ağva yolu üzerindeki Göksu köprüsünün ilave inşaatı olan betonarme işinin pazarlıkla eski müteahhide yaptırılması.
65. Akkaya (1989) Page: 68.
66. **Karl von Terzaghi** (1883–1963) an prominent geotechnical engineer, and geologist known as the "father of soil mechanics and geotechnical engineering". He lectured in the Higher School of Engineers (Mühendis Mekteb-i Alisi) in İstanbul and Robert College from 1916 to 1925.
67. Akkaya (1989) Page: 35.
68. Akkaya (1989) Page: 67.
69. Nafia (1933) Pages: 55, 56.
70. Bridges codes are: Irva (ACH-01), Koyunağılı (BRD-09), Silahtarağa (BWS-06), Sünnet (BWS-07), Manavgat (MTL-02) and Vezirhan (MTL-03).
71. Nafia (1933) Pages: 55, 56.
72. Aslanoğlu, I. (2001) *Erken Cumhuriyet Dönemi Mimarlığı 1923-1938*. ODTÜ Architecture Faculty Press. Ankara. Pages: 26–30, 92–99.
73. Sey, Y. (2003) *Türkiye Çimento Tarihi*. Tarih Vakfı. TÇMB. ÇMIS. İstanbul. Page: 32.
74. Örmecioğlu (2010).
75. **Fevzi Çakmak** (1876–1950) was a Turkish field marshal (Mareşal) and politician. He served as the Chief of General Staff from 1918 to 1919 and later the Minister of War of the Ottoman Empire in 1920. He later joined the provisional Government of the Grand National Assembly and became the Deputy Prime Minister, Minister of National Defence and later as the Prime Minister of Turkey from 1921 to 1922.
76. **Kazım Karabekir** (1882–1948) was a Turkish general and politician. He was the commander of the Eastern Army of the Ottoman Empire at the end of WWI and served as Speaker of the Grand National Assembly of Turkey before his death. He pledged with Mustafa Kemal to join the Turkish national movement and then took the command of the Eastern Front during the War of Independence. He founded the Progressive Republican Party in 1924. Then arrested and release with in connection with the İzmir conspiracy in 1926, but released. He entered again the assembly after 1938.
77. Bridges codes are: İsmetpaşa (Kömürhan) (ACH-09), Fevzipaşa (ACH-11), Karabekir (BWS-12), Çetinkaya (BWS-16) and Gazi (MTL-07).
78. Bridges are: Kömürhan (ACH-09), Aksu (BWS-10), Gediz (BWS-11), Borçka (MTL-04), Singeç (ACH-20), Sakarya (BWS-15), Çetinkaya (BWS-16), Gülüşkür (BWS-19) and Gazi (MTL-07).
79. İlter, İ. (1973) *Boğaz ve Haliç Geçişlerinin Tarihçesi*. Karayolları Genel Müdürlüğü Matbaası. Ankara
80. Bayındırlıkta 50 Yıl (1973) T. C. Bayındırlık Bakanlığı. Ankara
81. Walther, R. (1969) Spannbandbrücken: Vortrag. Zeitschrift: Schweizerische Bauzeitung Band (Jahr): 87 (1969) Heft 8. Retrieved from: http://doi.org/10.5169/seals-70602.
82. Herzog, M. (1966) Projekt einer Hängebrücke über den Bosporus. Zeitschrift: Schweizerische Bauzeitung Band (Jahr): 84 (1966) Heft 7. Retrieved from: http://doi.org/10.5169/seals-68840.
83. The sculpture designed by calligrapher Emin Barın. It is 1.5m high and writes "Maşallah" in Arabic letters in Kufic script.

84. **Mimar (Architect) Sinan** was the Royal Chief Architect-Engineer during the reign of Sultan Suleiman. Two of his remarkable bridges are Mağlova Bridge and Büyükçekmece Bridge.
85. It is a metaphor of dwarfs standing on the shoulders of giants and expresses the meaning of "discovering truth by building on previous discoveries". The most familiar and popular expression is attributed by Isaac Newton: "If I have seen further it is by standing on the shoulders of Giants." https://en.wikipedia.org/wiki/Standing_on_the_shoulders_of_giants.
    Griggs, F. E. (1994) On the Shoulders of Giants. *Journal of Professional Issues in Engineering Education and Practice*, 120:3, 254–264.
86. Martykanova, D. (2010) *Reconstructing Ottoman Engineers: Archaeology of a Profession (1789–1914)*. Front Cover. Darina Martykánová. Edizioni Plus, Pisa University Press. Pisa. Page: 180.
87. İnalcık, H. (1985) *Introduction to Ottoman Metrology. Studies in Ottoman Social and Economic History*. Variorum Reprints. London.
88. Engineer Kemal (1933) Cumhuriyet Devrinde Türk Mühendislerinin Köprücülük Faaliyeti. Arkitekt Magazine. No: 1933-09-10 (33-34) Pages: 283–290.
89. BCA-10-30-16-30-18-1-2/Kararlar Daire Başkanlığı (1928-) Belçika'da toplanacak Mekanik Jeneral, Beton-Betonarme ve Yol Kongrelerine temsilci gönderilmesi.
90. The formal letter of Ferruh Atav's resigning from Company. Retrieved from: https://www.bitmezat.com/en/product/1343343/sadik-diri-halit-kopr.
91. Turkey Engineering News (1964, 1 January) IMO. Ankara.
    The date Köprücü joined SAFERHA is not very clear in the reference documents. Örmecioğlu (2010) states the date as 1933, however a letter dated 1928 for the preliminary design of Kömürhan bridge has his name in the company title. On the other hand, Nafia descriptions name only Sadık and Ferruh as contractors for individual description of bridges until 1930.
92. Dibsiz Keson İmalinde Yeni Bir Keşif (1941) Arkitekt Magazine. No: 1941/42-03-04 (123-124) Pages: 63–68.
93. M.R.E. (1952, May 19) Köprü mü, Tünel mi? Interview with Halit Köprücü. Akşam. Page: 7.
94. BCA 230_9_16. 25.04.1952 dated Balıkesir ve Mameki köprüleri inşaat sürelerinin uzatılması.
    Letter is about the decisions and time extensions regarding the bridge projects which were contracted to SAHA company. As understood from the letter, SAHA applied for a liquidation agreement, and it has been approved.
95. Akkaya (1989) Page: 68.
96. Akkaya (1989) Page: 77.
97. Akkaya (1989) Page: 231.

Chapter 4

Bowstring Bridges

4.1 INTRODUCTION

The name bowstring derives from a bow-and-string analogy, in which the string holds the ends of the bow keeping the arch shape of the bow by resisting the tension, that is, it refers to the tie rod. At the onset of the use of this system, the name was "bow and string" bridges. Bow as a term was also used to describe an "arch" alone in bridge engineering, at least in some literature recorded during the 18th–19th centuries.

In this bridge type, the arch is above the deck level whilst the tie beam is connected structurally to the arch ends, thus restraining the thrust. The arch carries the deck through hangers. The deck provides a continuous surface for traffic and is formed by a slab, secondary beams and longitudinal beams. These longitudinal beams are mostly independent elements and are also called edge beams as they are located at the sides of the deck cross section, but in some cases, they can be the tie beam itself depending on the individual structural design of the bridge. Bridge structure is formed with at least two arches. When they are interconnected above the deck through lateral beams, these lateral beams are called wind beams.

Nowadays, the term "tied arch" is more commonly used instead of bowstring. In addition, the bowstring term has been applied rather indiscriminately, sometimes referring to a bridge that is not tied but a "through arch", in which the arch rise is above the deck. However, when applied to concrete bridges, it is more definitive as they were one of a special kind. This type was very popular for some time around the world for concrete design but has been all but abandoned. In most countries, the term "bowstring" has generally been used to refer to these bridges constructed during the era 1920–1940. In this book, the term will be used for concrete bridges unless it is stated otherwise.

Additionally, some bridges have been designed with this structural form, but with a very shallow rise and known as "bowstring girders". They have been fabricated with cast iron and mainly used for railway lines, even though these are called girders, and they were structurally arches with a relatively shallow rise of the arch (Figure 4.1).

Structurally, the span range for bowstring bridges typically starts from 20 m and over the years, progressively extended up to 42 m. In terms of span length, reinforced concrete beams have a limit of around 15 m to at most 20 m, for serviceability and practicality reasons. Therefore, bowstring bridges were very effective to provide solution beyond 20-m-long crossings for concrete bridge designs.

As a structural system, bowstring bridges were especially useful whenever ground conditions were not suitable to support the thrust from the typical arch bridge. By using

DOI: 10.1201/9781003175278-4

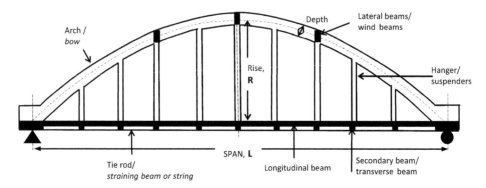

Figure 4.1 Bowstring bridge terminology (author).

ORIGINAL DESIGN FOR A FLOATING PIER AND APPROACH.

Figure 4.2 Brunel's "landing" bridge as self-standing bowstring from Beamish (1862).

the bowstring system, abutments and piers only carried the vertical loads from the super-structure and tie beams restrained the tension, that is thrust of the arch. This statement is true for the symmetrical dead load of the structure, whereas live loads exerted vertical loads and also moment depending on its point of application and shared with respect to relative stiffness between the arch and tie. Therefore, metal bowstring bridges were in some form of truss, that is triangular, and had diagonals between top and bottom chords. Instead, concrete arches with their robust cross section to carry loads allowed exclusion of the diagonals providing a moment of inertia with their structural depth.

The landing iron bridge designed by Brunel is a good example to show a bridge span independent from its supports (Figure 4.2).[1]

Another reason favouring the use of bowstring bridges was the full clearance available underneath, especially for railways and navigational rivers. The flat soffit of the bridge structure and the clear space beneath the bridge provided great advantages allowing as much clearance as possible not only for train passage and also for the navigation and the flow of water.

Last but not least, bowstring bridges have an independent system for each span; therefore, differential settlement between the substructure units does not create subsequent stresses in the superstructure, and these movements are independently tolerated within the span.

In terms of its origins, they can be regarded as an evolution from an arch design with the tie introduced to take care of the arch thrust and also be regarded as a truss type with tie as the lower chord and the arch as a curved chord. Furthermore, in locations where masonry arches were traditionally used for bridge structures, like in Europe, the former is more applicable and the bowstring type is more commonly used with suspenders only. Gauxholme Viaduct and high-level bridges are good representatives for this case. At the same time, the so-called bowstring girders are also used, like the Windsor bridge of Brunel in the UK.

Early bridges in the US were mostly constructed with timber and were primarily truss structures. Later, metal bridges continued similar design principles and had diagonals. When concrete came into use as a new material, the design developed from truss without diagonals. For example, conversions from metal bowstring bridges to concrete bowstring bridges were the utilization of concrete simply to stiffen and also protect the metal member within an encasement.

4.2 HISTORY OF BOWSTRING BRIDGES

The principle behind bowstring structures is quite ancient, and their use can be discerned in many situations. The early domed-structure buildings had iron ties to stabilize their arches. In bridge engineering, the centring of masonry arches was sometimes built with bowstring systems, which were probably a prelude to this system. On the other hand, the application of the bowstring system to girders with wrought iron rods, externally applied on tension sides of the section, was a forerunner to the prestressing technique in beam systems.[2]

The earliest document found that mentions the bowstring system is a book called "Machinae Novae" by Faustus Verantius,[3] a book written in 1595 for technical innovations involving bridges made from wood, stone and metal (Figure 4.3).

Another early document that mentions the bowstring term and explains the principle was Fulton[4] (1796) who used the arch together with a truss system in the USA. Fulton explains the system very clearly as follows (Figure 4.4):

> ... an arrangement of parts which, I conceive, would stand without abutments, this may be considered as a bow and string; which string, by keeping the bow bent, answers the purpose of abutments; all of the other braces being to preserve the bow and string in their proper situation by dividing the weight on the bow...

A year later, in 1797, James Jordan[5] was issued a patent for a bowstring bridge in the UK. His description involved all members except joint details. Tie beam described as "...horizontal ribs of bridge way, composed of well-seasoned timber, ...with cast iron plates, ..., the timber and the iron bolted together...". It seems to have inspired Burr in

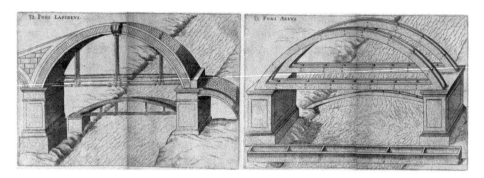

Figure 4.3 Left: "Pons Lapideus" (stone bridge) and right: "Pons Areus" (metal bridge) from Verantius (1595).

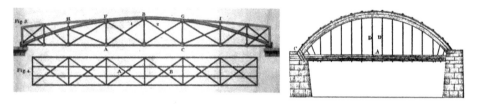

Figure 4.4 Left: Fulton's bowstring bridge (1796) and right: James Jordan patent (1797).

Figure 4.5 Sketches to show the system for bridge applications from Nasmyth (1796).

his Trenton bridge design, which is almost a precise copy of Jordan's patent drawings according to Griggs.[6]

Another early description of the system belongs to Alexander Nasmyth (1758–1840). As claimed by his son James Nasmyth, the bow-and-string bridge system was "invented" by his father in 1794. Sketches provided in the biography book[7] of the family show designs of metal bowstring bridges for a variety of spans (Figure 4.5).

The earliest practical use of the system was observed in timber bridges, especially in the USA and in mountainous parts of Europe. Some specific examples are the remarkable bridge of the Grubenmann brothers, in Wettingen dated 1764 in which the lower chord of the truss acted partly as a tie absorbing some of the thrust. Another more precise use of the system was the Rabiusabrücke bridge in Versam, which had demonstrated a structural form and proper arch and tie connection.[8] Later in the USA, Wernwag's Newhope bridge of 1813–1814 had a similar design, and the arch–tie connection was also obvious. It seems that the principle of a tied arch was understood, but the practical application was still not very common.

Similar applications were quickly made with the introduction of iron as a promising new material for construction. Railroad bridges demanded stronger materials, and also the flat soffit provided by bowstring type was needed for railway line envelopes passing under. As these bridges did not need a high rise like a concrete arch, they were called "bowstring girders". This was a truss system using cast or wrought iron at the top chord and wrought iron for bottom tie. These were very common with the first applications in the UK on the London-Birmingham Railway line in 1830. One example is the High-Level Bridge of Robert Stephenson in 1849, spanning the River Tyne between Newcastle and Gateshead in North East England.

Metal bowstring as a sole structural system spread as quickly as iron production improved to produce desirable cross sections and lengths. Then, in 1841, Squire Whipple patented a tied arch system and called it "iron truss bridge".

Although there is no definite date about where the system was first used in reinforced concrete bridges, there are some early bridges listed by different sources. The earliest example found by the author is the 1904 Avranches bridge crossing over the railway line in France. The bridge was built by Armand Considere, and it was a copy of a metal bowstring bridge constructed with reinforced concrete (Figure 4.6).[9],[10]

Considere tested his design with a 1/3 scale model of a 65.6 ft. (approximately 20 m)-long bridge. The test carried out in 1903 found that the 20-m model bridge showed no signs of failure at 85 tonnes which also includes 25 tonnes of dead load of the bridge when the load stayed 12 hours. The structure failed at 241 tonnes of the total load. Details of the test can be found in the appendices of the 1904-dated book on reinforced concrete.[11] The advantages of bowstring system are also described in this book in terms of their load-carrying behaviour; the loads are mainly transferred as direct loads with uniform intensity and the durability provided by the concrete. In addition, it mentioned that "…reinforced concrete enables the full resistance to be obtained from the metal, there being no deductions …for rivet holes, etc... (Figure 4.7)".

The earliest known two concrete examples found in America were both dated in 1909: the first one is the Sparkman bridge designed by Howard M. Jones in Nashville, Tennessee, and the second one is Benson bridge designed by German Immigrant E.A. Gast in Cincinnati, Ohio. Both bridges were designed without diagonals.[12]

Another example found in Canada is the first concrete truss bridge in Toronto, Ontario, designed by Frank Barber.[13]

All the above three bridges are currently standing today.

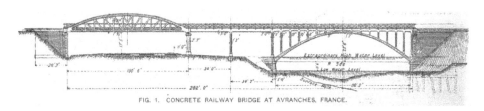

FIG. 1. CONCRETE RAILWAY BRIDGE AT AVRANCHES, FRANCE.

Figure 4.6 Concrete railway bridge at Avranches (Pont d'Ivr), France (Engineering News, 1909).

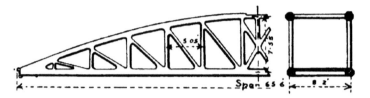

Figure 4.7 Details of Considere's bowstring test (Marsh, 1904).

FIG. 4. VIEW OF CONCRETE TRUSSES, SPARKMAN ST. BRIDGE. A Concrete Truss Bridge on the Middle Road, between the Counties of York and Peel, Ont.

Figure 4.8 Left: Howard Jones's concrete truss bridge from engineering news (1909) and right: middle road bridge from Canadian engineer (1909).

These pioneering bridges were the onset of a new era of building concrete truss bridges, and, in the USA and Canada, there are many surviving examples today, whereas in Europe, few remain. It is probable that World War II led to the destruction of many. Britain has a few examples which survived: one is the Milton Regis Railway bridge dated 1923, another is the Etive bridge dated 1932,[14] and the third is the Stow bridge dated 1926 (Figure 4.8).[15]

As reinforced concrete was a newly adopted structural material, these bridges received inevitable criticism such as: "...indefinite distribution of the stresses and the possibility of deterioration through impact and vibration...".[16] Criticisms were also related to the indeterminacy for the joints, secondary stresses imposed at the joints and also the unnecessary concrete weight used to cover the tension members. Another comment was the tendency for the development of hairline cracks in the concrete when a member is working under tension, such as in the case of ties and hangers. Later development addressed some of these issues, for example, to prevent hairline cracks in hangers, the concreting of them was left until after the main concrete-carrying members of the structure were completed.

4.3 SIGNIFICANT FEATURES OF BOWSTRING BRIDGES IN TURKEY

In Turkey, there are currently at least 19 bowstring bridges, still in good condition out of 24 of which, the author found that have been built from research. In Turkish, the term is almost directly adopted with a little revision as "baustring" and sometimes "bavstring", no specific Turkish name is applied for this type of bridge. It is probable that "u" or "v" was chosen to replace the English letter "w", which does not exist in the Turkish alphabet. Turkish "the tied arch" term is mostly used for metal bridges and called "gergili kemer" which translates to "strained arch".

One of the earliest plans for a bowstring bridge was found from Ottoman times, dated 1912 and signed by engineer A. Voudral. The crossing was in Bursa on the Orhaneli (Atranos) road over Kocasu river. The bridge plan was for a single-span bowstring type with a 30.4-m clear span between abutment walls.[17] As seen from drawings, the bridge was designed as a solution for crossing the gap when the previous masonry bridge collapsed. There is no further information found regarding the bridge (Figure 4.9).

Bowstring bridge construction in the early Republican era started with the Adagide bridge in 1925, and the last historical bridge found in the sources being Nif bridge dated 1950. Their design and evolution can be easily observed through the years. This progress can be most simply seen with the increasing span length. The earliest spans built in 1925 were 26 m and increased to 34 m for the Silahtarağa bridge in 1932. 34–36 m span length was the most common span length for all others with some exceptions. The last two bridges Nif and Melen had spans of over 40 m.

Another addition to the collection of bowstring bridges was found in İstanbul. Basmane bridge in Bakırköy had been constructed with this system, presumably around

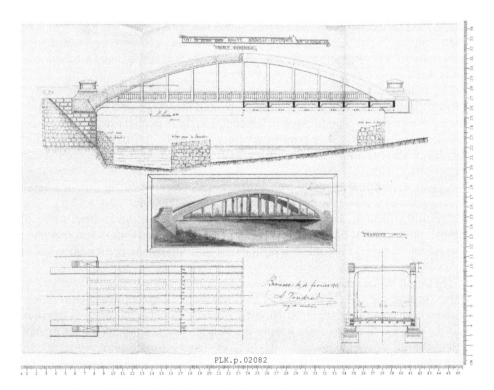

Figure 4.9 1912-dated Orhaneli bridge on the Bursa-Atranos road (BOA PLK-p-02082).

Figure 4.10 Sakarya bridge from upstream side (author).

1930. The arch of the bridge has a trussed form rather than solid cross section; therefore, it is described in Chapter 7 of this book.

Bowstring bridges have a distinctive aesthetic appearance. Although the appeal is debatable, these bridges are quite remarkable due to their prominent elevation over the road. Many bridges go unnoticed by road users as they stay below the road and do not have a chance to show themselves unless the road profile has a very significant curved contour. In this sense, bowstring bridges are very different as it is not possible to ignore them. Therefore, in most cases, these bridges easily become landscape icons (Figure 4.10).

The main structural parameter for an arch bridge is the ratio of rise to span. This parameter is given only for a few bridges in related literature, and the usual value is 7 m for a 35 m span, corresponding to 0.2, which is in the expected range of 0.16–0.30 for a bowstring arch bridge.[18] The rise of the bridge is usually limited by the navigation or vehicular clearances and roadway grades, whereas for bowstring type, the rise can be selected with complete freedom.

A greater rise will decrease the arch thrust, so decreasing the tension in a tie which in return reduces the axial stress from permanent and transient loads and the bending stress resulting from the variations in temperature. In opposition to these advantages, is the increase in height of the arch with increased length, and thus, reinforcement and height of the hangers, the total use of material and finally the construction cost will be greater.

The depth of the arch cross section is a governing value in design as it will determine the self-weight and the capacity of the section. The arch depth varies at the crown and in the springing. For example, Nazilli (Menderes) bridge has an arch depth of 50 cm at the crown and 62 cm at the springing. Values change again as the spans shorten and lengthen for the same bridge.

In bridge engineering, span and height above ground level of the bridge are parameters dependent on-site conditions and technical constraints, but the width of the bridge can be regarded as an independent value determined by the owner, mainly

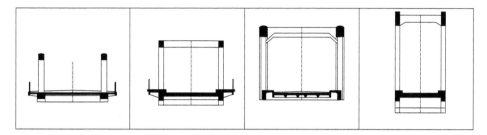

Figure 4.11 Sketch showing different cross sections from left to right: İncesu, Menderes, Silahtarağa and Aksu Bridge (author).

governed by traffic requirements. In regard to the bridge transverse section, different applications are observed. The earlier bridges had 100-cm-wide sidewalks on the outer sides of the arch, called "pedestrian step". This feature is observed in Gazi, Ankara, İncesu and Menderes bridges built between 1926 and 1931. In later bridges, the "pedestrian step" is replaced with sidewalks inside the arches, adjacent to the road, and its width was varying for each bridge with values of 60–100 cm. A minimum of 60-cm sidewalk next to the road also provided a "shoulder" to prevent the collision of vehicles with the hangers (Figure 4.11).

We can also conclude that there were no set parameters as dimensions were individually adopted for each project.

The design and details for lateral beams and cross bracings along the arch, also called wind beams, interconnect the arches above the deck level, improved over time. There were either none or only two beams in the older bridges, and this number is increasing up to four in the 1932-dated Silahtarağa bridge. Later on, this number interconnecting elements increased as span increased. Some also included K bracing, named like this as they resemble the letter K on plan view. This K bracing was applied to all bridges from 1939-dated Gülüşkür bridge onwards, and the number of transverse members increased to 5, 7 or 9.

The lateral bracing of the arches in the bowstring type is especially important to provide lateral stability to the structure. The number is directly proportional to the span, and four bracings are typically used where the span of the bridge is larger than 30 m. These members are placed as far as possible along the arch ribs starting from the crown of the arch down to where the clearance of the required cross section is achievable allowing the minimum headroom for the crossing vehicles (Figure 4.12).

Structurally, the arch and the tie beam carry the axial forces of compression and tension, respectively. The moment carried by these structural elements is shared as per the stiffness of each section. Public Works Authority/Nafia[19] (1933)[20] report specifically mentions the structural design intent of the elements for some bridges. For example, the Gazi bridge is described as carrying axial load and moment by its arch and only axial load by the tie, whereas the Adagide bridge does not carry the moment within the arch. This might be explained by the Adagide arch not having reinforcement within the concrete section. Then, later bridges, Ankara and Nazilli, mentioned that the arch and tie beam was to carry both axial load and moment.

Figure 4.12 Akçaağıl bridge view (author).

Another important structural and architectural feature of these bridges is the connection between tie beams, transverse deck beams and deck slab connections. They are detailed quite differently from each other and had a remarkable impact on the visual appearance of the bridge. Some of these details are shown in the figure below to provide an insight over the structural design variations (Figure 4.13).

The bridge approaches and entrances are also important in defining the bridge's character. Almost all bridges had a specific detailing for the approach parapets. The deck has a rail between the hangers for all bowstring bridges, and this rail continues after the ends of the arch in many bridges. This can still be observed in Sakarya and Yalakdere bridges. Some bridges had solid parapets at their approaches. Horasan, Çetinkaya and Gülüşkür bridges were some examples of this feature, and their parapets were also widened to have a clearer entrance to the bridge.

İncesu, Keloğlu and Gazi bridges also had well-detailed bollard-like posts at the entrances over the arch and tie connections.

Silahtarağa, Sünnet and Ankara bridges had tall lighting poles featuring at each entrance. These can be classified as metropolitan bridges; therefore, this treatment is understandable, as they most probably had stronger architectural design input during the concept development and design of the structure.

Construction of bowstring bridges is more challenging compared to other bridge types like deck arch or beam. This results from the complexity and staged construction of many different structural elements and joints in a progressive manner.

As a special construction application utilized in bowstring bridges, pretensioning was applied to the tension members of the bridge during construction. In the first stage, the whole arch, deck and deck beams were concreted. Hangers and tie beams are left to the second stage, and the middle third of the deck is also left to the second stage.

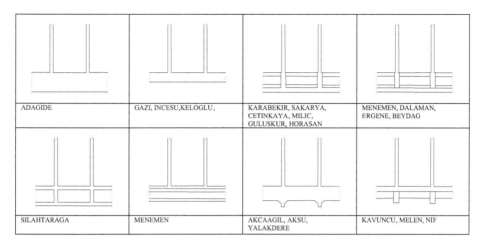

ADAGIDE	GAZI, INCESU,KELOGLU,	KARABEKIR, SAKARYA, CETINKAYA, MILIC, GULUSKUR, HORASAN	MENEMEN, DALAMAN, ERGENE, BEYDAG
SILAHTARAGA	MENEMEN	AKCAAGIL, AKSU, YALAKDERE	KAVUNCU, MELEN, NIF

Figure 4.13 Connection details for deck–tie–hanger from side view (author).

Pretensioning was thus naturally applied by the self-weight of the already concreted sections. At times, additional surcharge loads are applied where needed. It was found that this application was carried out for the first time in Turkey during the construction of Silahtarağa bridge. Gediz, Karabekir and Çetinkaya are the other structures that this method is known to be utilized.

4.4   EARLY REPUBLICAN BOWSTRING BRIDGES IN TURKEY

Concrete bowstring bridges in Turkey were not built in great numbers, as this research found a rather low number. Therefore, the few surviving bridges that retain their original character are of significant value in terms of their heritage. These defining features would include top chord, tie and hangers. In addition, in many cases, the railings and elements at the entrances may also be categorized as character-defining features of these bridges. The railing details are also considerable, especially when these are adopted between hangers, through the expansion joints, at approaches in order to protect the arch member and detailed provide a good transition from road level to the rising arch (Figure 4.14).

There are a total of 24 concrete-tied arch bridges found and described in this chapter. Amongst them, 19 are currently standing and in good condition representing the unique collection of their type, that is bowstring bridges in Turkey. Most of them are in relatively good condition and eligible to be regarded as heritage bridges, as they are the last stand of a small group of specific bridge types from their era. The full list of these bridges is given below with some information such as name, completion date, location, obstacle crossed (mainly river), number of spans, length of its main span, current condition, designer or contractor, and coordinates (Table 4.1).

Some bridges which did not survive to this day are also included in the list since they provide a significant contribution to the history. Some details are explained below:

Figure 4.14 Bowstring bridge locations shown on Google map (author).

Adagide bridge was damaged by the flood in 1979 and subsequently collapsed.

Sünnet bridge was demolished in the 1970s during the Golden Horn Estuary Restoration projects.

Keloğlu bridge is documented as built; however, it could not be found and therefore is presumably no longer standing.

Akçaağıl bridge was flooded by the lake of a dam. When the water level drops during summer, this bridge becomes visible and can be utilized for crossing.

Ergene bridge is known to have been built; however, it could not be found and is presumably no longer standing.

Gülüşkür bridge was submerged within a dam lake.

Çağdırış bridge exists in KGM lists; however, no other record has been found about this structure.

BWS-01: Adagide bridge

Adagide was the first reinforced concrete bridge of Turkey and the second bridge built by the Republican Public Works. It was designed by Mehmet Galip Sınap. Built over the Küçük Menderes (Little Meander) river on the Ödemiş-Adagide road in İzmir, it had three spans, each of which was 26 m. The construction of the bridge started in 1924 and was completed in 1925 (Figure 4.15).

The unique features of this bridge were the high parapet walls at the arch and tie beam joints. These walls were higher than usual parapets.

The designer, Mehmet Galip Sınap (1888–1962), graduated from Ecole des Ponts et Chaussées[21] (Paris) in 1913. Sınap worked as a teacher between 1919 and 1922 at the Engineering School in İstanbul. He later formed a company based in İzmir, which was in the same region as the bridge site. Sınap was the designer of one of the first reinforced concrete buildings in Turkey named Kardıcalı Han (1928) in İzmir (Figure 4.16).

Unfortunately, the bridge did not survive to current times. According to the local newspaper[22], "In December 1979, the Menderes River overflowed from its bed when heavy rains flooded the plain. Over the years, sand taken from the west side of the bridge weakened the river bed and the bridge collapsed transversely on its side".

Table 4.1 List of Bowstring road Bridges in Turkey (author)

BWS No	Name	Comp. Date	Location	River	No of Spans	Main Span Length (m)	Status	Designer (d.)/contractor (c.)	Coordinates
1	Adagide	1925	Ovakent İzmir	Küçük Menderes	3	26	Collapsed	d. Galip Sinap	38.16742, 28.01343
2	Gazi Orman Çiftligi	1926	Ankara	Ankara Çayi	1	25	Standing	c. Tahsin Bey	39.94931, 32.79207
3	Ankara Menderes	1929	Zonguldak	Büzülmez	1	26	Standing	SA FER	41.45016, 31.7919
4	Nazilli	1931	Aydin	Menderes	3	28	Standing	SA FER	37.87579, 28.32768
5	Incesu Atav	1931	Çanakkale	Incesu	1	25	Standing	SA FER	40.02414, 27.03936
6	Silahtarağa	1932	Istanbul	Alibey	1	34	Standing	d. Naşit Arikan	41.06846, 28.94368
7	Sünnet Fil	1932	Istanbul	Kağithane	1	34	Demolished	d. Naşit Arikan	41.06461, 28.94816
8	Keloğlu	1932	Mersin	Keloğlu	1	25	Incomplete Information	d. Vehbi	36.92737, 34.94242
9	Akçaağil	1933	Sivas	Kelkit	2	35	Half Submerged	c. Behcet	40.229, 38.04607
10	Aksu	1933	Antalya	Aksu	3	42	Standing	Saferha	36.94361, 30.89674
11	Menemen Gediz	1935	İzmir	Gediz	5	31.33	Standing	Ankara inşaat	38.64916, 27.05161
12	Karabekir	1935	Ankara	Delice	1	32	Standing	Muhtar Arbatli	40.00202, 34.08105
13	Dalaman Atatürk	1936	Muğla	Dalaman	3	35	Standing	Saferha	36.83427, 28.79452
14	Yalakdere	1936	İzmit	Yalakdere	1	22.5	In use	Saferha	40.69204, 29.50234
15	Sakarya	1937	Bolu	Sakarya	3	35	Standing	Saferha	40.69204, 29.50234
16	Çetinkaya Bafra	1937	Samsun	Kizilirmak	7	35	Standing	c. Mahmut Huseyin Mustafa Resit	41.56714, 35.88005
17	Ergene	1937	Tekirdağ	Ergene	4	31.33	Incomplete Information	Saferha	41.21633, 27.53054
18	Miliç	1939	Samsun	Miliç	1	31.8	In use	Hasan Selahattin Berkeman	41.17595, 37.04635
19	Gülüşkür	1939	Elaziğ	Murat	5	36	Submerged	Aral	38.64327, 39.71752
20	Horasan	1940	Erzurum	Aras	3	36	Standing	Kemal Gençspor	40.03549, 42.18522
21	Kavuncu	1945	Eskişehir	Porsuk	1	30	Standing		39.41391, 31.97857
	Çağdiriş	1946	Trabzon	-	-	-	Incomplete Information		
22	Beydağ	1948	İzmir	Menderes	1	35	Standing	STFA	38.10291, 28.20363
23	Melen	1948	Düzce	Melen	1	42	Standing	STFA	40.86028, 30.9824
24	Nif	1950	Manisa	Nif	1	42	Standing		38.50508, 27.61699

Figure 4.15 **Adagide bridge drawing (author).**

Figure 4.16 **Adagide bridge from Nafia album (İBB Atatürk Library).**

BWS-02: Gazi (Orman Çiftliği) bridge

Gazi[23] bridge was constructed to cross over Ankara Creek between Gazi Railway station and Ankara-İstanbul road. It is also known as Orman Çiftliği bridge (Atatürk Forest Farm) for the nearby Atatürk Forest Farm and Zoo, a recreational farming area which houses a zoo established in 1925. A bowstring-type bridge with a 25-m span was chosen since the ground conditions were poor, and a flat bridge profile was desired. This bridge has sidewalks outside the arches and a small bollard on top of the arch and tie beam connection.

The bridge construction was tendered in August 1925 and opened in April 1926.

The bridge was repaired in 2000 by KGM. Their report provides bridge dimensions of 28.1 m length and 9.6 m width.

The contractor of the bridge is given as Tahsin Bey in KGM documents (Figure 4.17).

BWS-03: Ankara bridge

This bridge is at the start of the Zonguldak-Devrek road and adorns the entrance to the city Zonguldak, or at least, it did when it was built.

Figure 4.17 Gazi bridge from Nafia album (İBB Atatürk Library).

The bridge features original high columns at both entrances for the purpose of lighting. The pedestals for these columns are well detailed. The bridge span is 26 m with a width of 8.3 m between the parapets. The clear width between kerbs is 4.8 m, 20 cm between kerb edge to arch and an extra 105-cm sidewalk placed on the outer side of the hangers (Figure 4.18).

The bridge was tendered on 01 August 1928 and opened on 19 March 1929.

The ground conditions were very poor. Therefore, 34 and 64 concrete piles were used on Devrek and Zonguldak sides, respectively.

Ankara bridge was repaired by Zonguldak Municipality around 2014 as the bridge was under "danger of collapse" as stated in a newspaper.[24] During this repair, the damaged lateral beam was also repaired. In June 2014, whilst the bridge was under renovation, a strong flood occurred and the temporary metal scaffolding was washed away by floodwaters. A pedestrian bridge nearby collapsed, but fortunately, the Ankara bridge survived the flood.

The same lateral beam which had previously been repaired was again hit in 2020, and similar but less severe damage can be seen at the same location. The bridge is still open to traffic (Figure 4.19).

BWS-04: Nazilli (Menderes) bridge

This is a three-span bowstring bridge crossing over the Menderes (Meander) river on the Nazilli-Bozdoğan road. The spans are 22, 28 and 22 m. The clear width between the arches is 5.2 m, and width of the road is 4.80 m and. Pedestrian sidewalks are outside the arches with 100 cm width on each side.

Transverse beams are spaced at 2 m. The depth of the section of the arch is 50 cm at the crown and 62 cm at springing for end spans, and 60 cm at the crown and 75 cm at

Figure 4.18 Ankara bridge (author's collection).

Figure 4.19 Ankara bridge from Nafia album (İBB Atatürk Library).

the springing for the middle spans. The width of the arch section is 50 cm for all three spans (Figure 4.20).

Two lateral beams for wind loading were provided for the middle span, only one of which remains today. The end spans had none.

Along with Menderes bridge, Kalabaka and Dalama bridges, the construction contract dated 26 November 1929, were awarded to Sadık Diri and Ferruh Atav (SAFER). The construction of the bridge was completed on 10 August 1931 (Figure 4.21).

The main arterial road moved to a new bridge built in the 1950s during the Bozdoğan dam construction; however, the bridge continues to serve local traffic.

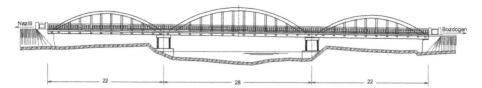

Figure 4.20 Nazilli bridge sketched by author (author).

Figure 4.21 Nazilli bridge from Nafia album (İBB Atatürk Library).

BWS-05: İncesu (Atav) bridge

İncesu bridge was on the Balya-Çanakkale road which was the first major road project of the New Republic government. Therefore, the inscription of the bridge states: "The first through bridge which was ordered by Atatürk to engineer Ferruh Atav". Indeed, Atatürk, the leader and founder of modern Turkey, signed all the governmental decisions, including major road projects as the head of the government (Figure 4.22).

Ferruh Atav, an engineer educated in Turkey, who also had training in Germany, was one of the founders of SAFERHA company which was one of the leading companies of that time. The bridge is also known by his name as "Atav Bridge". Balya-Çanakkale road, completed in 1933, was the first major road with asphalt in Turkey. The road and all the infrastructure were later declared as national heritage (Figure 4.23).

The bridge is crossing over the İncesu creek with a 25-m span. İncesu was known to flood regularly. For this reason, previous bridges built over the stream were destroyed due to the scour of piers. Therefore, pier construction in the stream was avoided, and the bowstring type of bridge was the preferred option.

The width of the bridge between the arches is 6 m, of which 5.60 m is the effective width of the road. The sidewalks are each 100 cm wide and located outside the arches. The bridge width is given as 9.4 m in total.

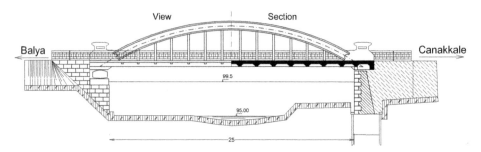

Figure 4.22 İncesu bridge (author).

Figure 4.23 İncesu bridge from Nafia album (İBB Atatürk Library).

There are no longitudinal beams other than the tie beams in the deck. The deck is formed as a continuous slab supported by transverse beams which have regular spacing at 1.7 m.

Bridge construction was awarded to SAFER with the contract dated 29 July 1930 and completed a year later.

BWS-06 and BWS-07: Silahtarağa and Sünnet (Fil) bridges

There were two bridges with the same design constructed in 1932. One is Silahtarağa bridge on the Kağıthane river, and the other was Sünnet bridge on Alibey river. Both rivers merge into one and form the north of the Golden Horn (Haliç) in İstanbul.

The author visited Silahtarağa bridge in 2016; however, Sünnet bridge could not be found in its presumed location. It was likely demolished in the 1970s during the Golden Horn Estuary Restoration projects.

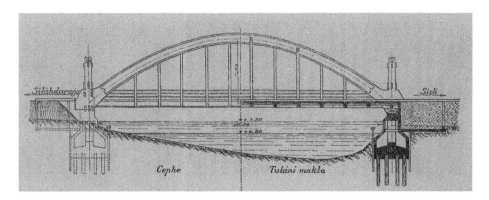

Figure 4.24 Silahtarağa bridge from Nafia album (İBB Atatürk Library).

Silahtarağa bridge is also known as Fil bridge after the previous bridge which was named like this in this location (Figure 4.24).

The span of the bridge is 34 m with a 7-m rise. 34-m span length was a considerable increase from the previous span ranges of 25–28 m. The bridge deck finish was made with tar, a protection layer was laid on concrete lastly and concrete blocks were used as pavement. The clear width was 5.30 m with 60-cm-wide kerbs on both sides. The total width of the bridge is 6.5 m.

The arch section is 70 cm wide with 80 cm depth at the crown, progressively increasing to 100 cm at the springing. Four lateral beams were provided to resist wind and transverse forces.

The slab rests on the six longitudinal beams and another two tie beams, which connect with 14 transverse beams at hanger locations. This system constitutes complete grillage for the bridge deck, separating the deck into equal sections. Transverse beams and tie beams are connected to the arch by hangers of 50 × 20 cm.

Ground conditions were extremely weak "soft and rotten" at the bridge locations. This was investigated during the construction of the nearby Silahtarağa Power Station.[25] Further investigations were needed with boreholes, and a timber test pile was driven to 23 m. These studies showed that the weak soil extended for at least 20 m below ground. Therefore, the foundation had to be laid on longer piles (Figure 4.25).

A bridge type that would exert horizontal forces at the substructure was discarded as an option, and a bowstring type was chosen in order to transfer mostly vertical forces.

Silahtarağa bridge was designed with an innovative method: in order to reduce the loading exerted on the piles, the abutments are detailed like a hollow box and a superstructure supported on bearings which sit on 90 × 90 cm reinforced concrete columns. Columns are then isolated from the caisson and act independently. Columns directly transfer vertical loads to the foundation, thus avoiding any horizontal loads on the piling system.

An engineer named Naşit Arıkan working at Nafia allegedly "made" the bridge in his short portfolio.[26] He was indeed a graduate engineer at this time, and probably he was undertaking his compulsory work in Nafia.

Figure 4.25 Silahtarağa bridge details (author).

In the Nafia (1933) report, the cost of the bridge is described, and also comments are provided on the changed management system upon the new rules published in 1929. With the new system, Nafia resumes the management of road and bridge construction again instead of local government authorities in the district.

As narrated from the Nafia (1933), "...province had a preliminary design of beam solution...". The work had been tendered for 237.344 liras for the construction of two bridges, excluding the road works at the approaches. However, the tender was not approved, and the project was re-tendered with a bowstring bridge system. According to the new project, the total cost of both bridges was 122.917 liras including the additional construction of the approach road works.

"This big difference between the two independent designs meant halving the cost of the bridge without comprising the design quality. This clearly shows the optimum benefit of the principles constituted by the Roads and Bridges Law[27] and especially the design and study of the bridges by experienced engineers".

Nafia also made some improvements in the tendering system as explained as follows:

According to the specifications, contractors were allowed to use one of the three types of the piles listed below as long as there will be no cost implications:

1. Timber circular piles, 2. Timber square pile, 3. Reinforced concrete pile

The most critical and expensive part of the construction was the piling. Providing flexibility in choosing the type of pile for contractors increased the number of submission for tender.

Figure 4.26 Sünnet bridge from Nafia album (İBB Atatürk Library).

The progress of the construction of bridges can be traced from newspapers. Some of the articles[28] are shown in Figure 4.26 below. Another paper from 29 June 1932[29] was reporting that even though Sünnet bridge construction finished, the sidewalks were not completed and therefore bridge was still not open to traffic. It also mentioned that during this time, the timber bridge alongside the new bridge remained in use for vehicle traffic (Figure 4.27).

Construction of the bridge was carried out with the cantilever method. As the ground conditions were poor, it was not possible to pile into the ground to support the centring. Therefore, in the first stage, the arches were poured with timber centring supported on abutments. Then decking, ties and hangers were constructed. Pretensioning was applied to the ties and hangers as well.

The project started with 25 August 1930-dated contract and was completed on 25 July 1932.

BWS-08: Keloğlu bridge

This 25-m span bridge built over Keloğlu stream on the Tarsus-Adana road in Mersin. The bridge is known to be built; however, it could not be found and is presumably no longer standing. The construction of Keloğlu bridge utilized an old method of using soil fill as scaffolding. The bridge was crossing an artificial channel, so probably a considerable part of the fill which allowed the construction was pre-existing soil. The methodology is currently known as "top-down" construction in today's engineering

Figure 4.27 Article from newspapers; left: "tender announcement of two bridges" (01 August 1930), right: "Sünnet bridge construction will soon be completed" (08 December 1931).

terms. In this method, the ultimate structure is built on the existing ground surface, and excavation is subsequently undertaken (Figure 4.28).

The superstructure utilized the same design as İncesu bridge. However, no sidewalks were built outside the arches.

The bridge takes its name from the stream which is dry most of the year but often floods seasonally during the rainy period. The local ground consists of strong clay. However, the floods progressively erode the river bed and banks. Therefore, various bridges that were built before were collapsed by scour around its pier. A bowstring system was preferred, because there was not enough headroom under the road level to construct a conventional arch.

Channel Construction: The stream's gradient was very steep; thus, it was necessary to make a new channel to regulate the river flow. The foundations of the abutment were built in the dry and rested on a hard soil layer. The foundations were built only 1 m lower than the lowest water level of the stream. However, both the bottom and the sides of the stream were lined with reinforced concrete at the bridge location. Shoring thickness is 10 cm with spaced joints at 3 m to prevent cracking.

The bridge was tendered on 4 December 1930 and finished on 30 December 1932.

BWS-09: Akçaağıl bridge

Akçaağıl bridge provides crossing over Kelkit river which has a long and deep valley, separating local communities. Shallow rivers can provide many alternatives for crossings; however, very deep gorges need to be crossed with bridges. Therefore, a permanent bridge was deemed necessary, as the previous timber bridge was swept away by a seasonal flood (Figure 4.29).

This bridge has two spans each of 35 m long. Suşehri side is resting on rock. The foundation on the Koyulhisar side was constructed using reinforced concrete caisson.

Figure 4.28 Keloğlu bridge from Nafia album (İBB Atatürk Library).

Figure 4.29 Akçaağıl bridge (author).

The pier foundation was 5.5 m deep. As the foundations were excavated relatively deep, a large amount of dewatering was needed. The work undertaken for the construction of foundation and the remote location of this bridge created a considerable challenge.

The width of the bridge is 4 m between the arches. The arch cross section dimension at the crown is 60 × 95 cm.

The first tender announcement of the bridge was found in a 19 June 1930-dated newspaper[30] describing the bridge as masonry bridge. This tender of the bridge was first awarded to SAFERHA on 27 July 1930. However, construction did not start, and later, a new tender was opened. The second tender announcement was made for a reinforced concrete bridge in a 20 August 1931-dated newspaper[31] and was awarded to engineer Vehbi and contractor Behçet on 2 September 1931 (Figure 4.30).

Strong winds through the valley led to the arches being interconnected by six lateral beams. The decking was made of concrete. The bridge was completed on 15 December 1933 (Figure 4.31).

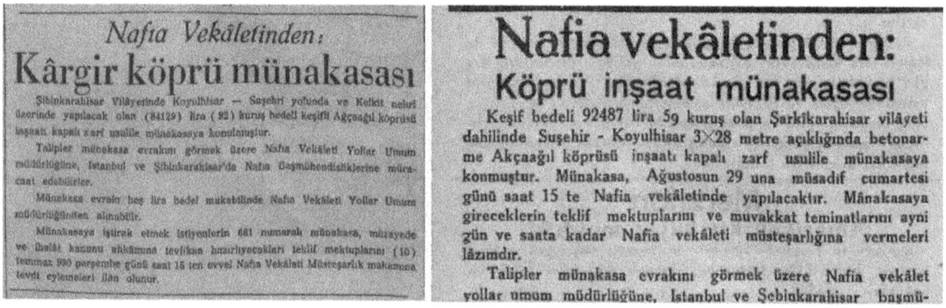

Nafıa Vekâletinden:

Kârgir köprü münakasası

Şibinkarahisar Vilâyetinde Koyulhisar — Suşehri yolunda ve Kelkit nehri üzerinde yapılacak olan (84129) lira (92) kuruş bedeli keşifli Ağçaağıl köprüsü inşaatı kapalı zarf usulile münakasaya konulmuştur.

Talipler münakasa evrakını görmek üzere Nafia Vekâleti Yollar Umum müdürlüğüne, İstanbul ve Şibinkarahisar'da Nafia Başmühendisliklerine müracaat edebilirler.

Münakasa evrakı beş lira bedel mukabilinde Nafia Vekâleti Yollar Umum müdürlüğünden alınabilir.

Münakasaya iştirak etmek istiyenlerin 661 numaralı münakasa, müzayede ve ihalât kanunu ahkâmına tevfikan hazırlıyacakları teklif mektuplarını (10) temmuz 930 perşembe günü saat 15 ten evvel Nafia Vekâleti Müsteşarlık makamına tevdi eylemeleri ilân olunur.

Nafia vekâletinden:

Köprü inşaat münakasası

Keşif bedeli 92487 lira 59 kuruş olan Şarkîkarahisar vilâyeti dahilinde Suşehir - Koyulhisar 3×28 metre açıklığında betonarme Akçaağıl köprüsü inşaatı kapalı zarf usulile münakasaya konmuştur. Münakasa, Ağustosun 29 una müsadif cumartesi günü saat 15 te Nafia vekâletinde yapılacaktır. Münakasaya gireceklerin teklif mektuplarını ve muvakkat teminatlarını ayni gün ve saata kadar Nafia vekâleti müsteşarlığına vermeleri lâzımdır.

Talipler münakasa evrakını görmek üzere Nafia vekâlet yollar umum müdürlüğüne, İstanbul ve Şebinkarahisar başmü-

Figure 4.30 Left: first tender announcement for bridge dated 19 June 1930, right: second tender announcement for bridge dated 20 August 1931.

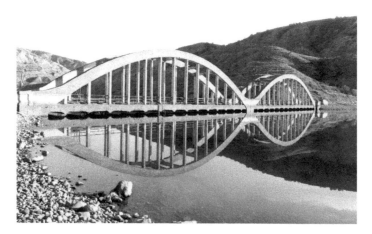

Figure 4.31 Akçaağıl bridge (author).

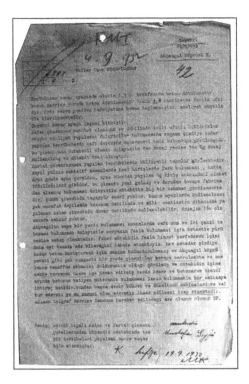

Figure 4.32 Akçaağıl bridge controller's letter reporting on-site activities (KGM archives).

Örmecioğlu[32] provides a copy of the letter dated 29 August 1932 written by the authority "controller" reporting on the construction of the bridge. This letter provides useful information to understand the condition of the time and is transcribed below (Figures 4.32):

To Roads General Directorate,

On the Koyulhisar side abutment, 1.5m deep concrete was poured with tremie[33]. On top of that, concrete will be poured in the dry. Caisson is horizontal with +/-0.80cm deviation. The pier caisson excavation has just begun. The operation is progressing with difficulty.

The construction of the Suşehri abutment is finished.

Even though Aslan cement is damp and contains large and small water particles – due to transportation in winter – it has been observed that it has had sufficient strength in the tests performed so far. Due to the dampness of this cement, 300 doses[34] were used instead of 200 doses and the contractor has agreed.

In the tests carried out on Kartal cement, full compliance was not observed. It is not only the difference between beams manufactured at different times but also the ratio between 3 beams made from the same sack on the same day which is 726:1206:1422. Since this cement has just arrived and was taken directly from the factory, no fault is

Figure 4.33 Akçaağıl bridge (author).

seen in the contractor, because no change was observed in cement. Although it is un-
likely to be used at piers, it can be used inside the caisson and inside a 50-cm-thick wall
to be constructed with Aslan cement. It is acceptable for the dosage to be 160.

The fact that Akçaağıl is located somewhere remote, there is a layer of sand and
coarse gravel in the caissons and excessive wind causes the work to go slow. However, I guess that the contractor can organize his job better by putting in more effort.
Although there is no machine for mixing concrete until now, and the cement is blowing
in the air and caused workers' throats to fill with it because of the heavy wind, the work
was continued to prevent the interruption to the contractor's schedule. But, the concrete for the arch and remaining structure has to be done with homogeneous concrete,
and necessary tools such as machines and sieves must be prepared.

As per the received telegram, the transfer to Samsun will start, E.F.

Footnote: 5 kilos each have been taken from the existing closed Aslan and Kartal
cement bags for strength experiments at the Engineer School.

Engineer Mustafa Seyfi [Tonga]

Akçaağıl bridge is in the lake of a dam; nowadays, the water level drops in the summer
so that bridge becomes visible and even provides access. The author visited this bridge
in 2020, the water level was just below the deck and the signs of deeper water levels
were traced from the colour difference in the structure. The bridge was being used by
people fishing and also a bride and groom for their wedding photography (Figure 4.33).

BWS-10: Aksu bridge

The Aksu river had a timber bridge that burned during a fire in 1914. A replacement timber structure was constructed supported on the remaining stone substructure. However, this one also burnt down in September 1930. Therefore, a more durable bridge was deemed necessary.

The order to construct the Aksu bridge was signed in 1930 after the previous existing timber bridge placed there was consumed by fire. This decision[35] was indeed an addendum to an existing list of bridges to be constructed with high priority. Aksu bridge joined this priority list as it was vital for the transportation between four nearby towns (Figure 4.34).

The new bridge also named Aksu bridge was another significant bridge with a span of 42 m. However, its main bowstring span was cantilevered to the adjacent spans and thus formed a cantilever-type bridge. Cantilever bridge was also called "Gerber beam/bridge" from the inventor, engineer Heinrich Gerber, who patented this system in 1866. In its structural concept, this bridge used a hybrid design (Figure 4.35).

In order to keep the road profile level in the approaches and considering that the ground capacity would not offer much strength, an arch bridge did not seem a feasible option. A bridge with steel superstructure could have been used to cross this span, but steel was not a very common material at that time and would have needed to be imported. Also, the bridge would need to be tendered in two separate packages for steel and concrete substructure separately. Therefore, it was decided to design the whole bridge with concrete as a bowstring type. An arch span would have required the piers to be considerably thick, as the piers were already tall, and this meant that the foundations need to be bigger increasing the cost. This problem was solved by cantilevering the ends of the arch beyond its piers. "In this way, the appearance of the bridge would improve, and the dimensions of the main span arch would also reduce..." as Nafia (1933) reported.

The bridge was designed with 18.90-m side spans next to 42-m main central span and 15.60-m end spans next to the abutments.

Figure 4.34 Aksu bridge on the Antalya-Serik road (author).

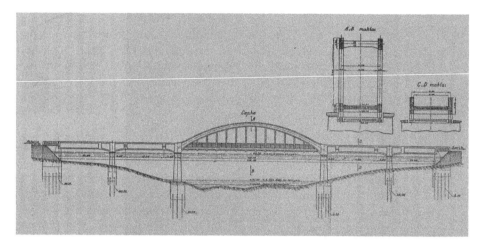

Figure 4.35 Aksu bridge section from Nafia album (1933).

Whilst removing the existing masonry abutments of the previous timber bridge, it was found that they were deeper than initially thought. At the same time, it was realized that it would be difficult to undertake piling works without removing the old foundations. Therefore, in order to avoid an increase in the cost, the end span on Serik side was shortened by moving the abutment and reduced to 13.50 m.

The pier widths are 1.5 m for those supporting the main span and 1.0 m for the others. Piers have a taper of 1/20 with a narrow width at the top. The reinforced concrete beam in transverse direction (diapragm) is provided on top of the piers to spread the load along the pier width.

The slab is supported by equally spaced transverse beams at 2.8 m. There is no longitudinal beam, and the transverse beams are directly supported on the tie beam of the arch.

The arch is 50 cm wide and 120 cm deep at the crown. The Gerber beam is 11.20 m in length and is supported by 3.85-m cantilever extensions from the piers. The beams are in line with the tie beam and align with the parapets of the bridge.

The bridge width is 4 m between arches with 30-cm-wide kerbs provided inside the arch for pedestrian paths and also to provide a safe envelope against a vehicle collision with the arch.

In order to ease the construction, the requirement of pretensioning the reinforcements was eliminated, and the formwork was removed soon after pouring the concrete of the whole span. Therefore, the reinforcement was over-designed to carry up to 1000 kg/cm² for tie beams and 900 kg/cm² for the slab for the ultimate load combinations.

Construction started on 2 April 1932 and finished on 29 March 1933. The bridge was opened by Governor Zeynel Abidin (Özmen) and joined with local notables with a huge crowd attendance on 21 April 1933 (Figure 4.36).[36]

Feyzi Akkaya[37] provides the following information[38]:

Aksu Bridge was one of the beautiful bridges of its time, as a replacement for a burned wooden bridge. This bridge is still standing, but abandoned with the new

Figure 4.36 **Aksu bridge (FATEV archive).**

road built next to it since the bridge was narrow. If you wish, you can still cross over it...

As my first work in Ankara, I was given the check of the structural calculations for this bridge by our Chief Kemal (Hayırlıoğlu)[39] Bey...

Aksu Bridge in Antalya was completed and its acceptance was ensured. Mithat Kasım and I constitute the delegation...

On a feast day, we moved by train, accompanied by Sadık Diri, one of the contractors of the bridge. We entered Afyon in the snow at 2 am in the early morning...

Loneliness and empty lands were predominant along the 14 hours of travel, in the road and the villages... Antalya was a quiet, calm, daisy-like city that was not filled with human silos like today. With our arrival, 12 engineers gathered in Antalya. Unprecedented thing!... Our photos[40] were taken...

...The Wisdom of God! A wooden bridge always burned a few months later after its construction... Later, by chance, the guilty were caught by the Nafia leadership... He was trying to break a piece of concrete with a wooden pattern using his axe. In his statement at the court, he declared that he was wondering if the bridge would burn or not... He was sentenced to two years.

BWS-11: Menemen (Gediz) bridge

The Gediz (Hermos) river is the second-longest river in Anatolia flowing into the Aegean Sea in Turkey (Figure 4.37). It has been an obstacle for roads to cross to reach İzmir (Smyrna) where a harbour always existed. Gediz had a timber bridge built during the First World War by the military. This bridge was on the Gediz river in İzmir-Bergama road. However, it was obsolete, and constant repairs were needed. The bridge had to be substantially rebuilt. Even though it was on top of the priority list, the repair works were not started due to the shortage in budget. Eventually, the İzmir

Figure 4.37 Gediz bridge (author).

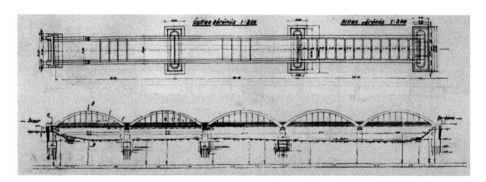

Figure 4.38 Gediz bridge (Nafia Works Bulletin Year: 1935/11).

Province took on a share of the expense, thus allowing the construction to be commenced (Figure 4.38).

The contract was awarded to a contractor named Hayri Kayadelen. However, he handed over the work to another company "İnşaat İdari Fenniyesi", Mehmet Galip and Fescizade İbrahim with a contract signed on 21 February 1933.

Two proposals were prepared, one with under deck arches with six spans of 30 m each and the other using a bowstring concept having six spans of 31.33 m each. The final choice was to be made once the ground conditions were determined on-site. During further investigation, one of the spans was removed by shifting the bridge location.

Thus, the bridge consists of five spans of 31.33 m. The section of the arches is 65 cm wide and 90 cm deep at the crown (Figure 4.39).

A development of Pretensioning: The structural parts, which ultimately work under tension; parts of deck, ties and hangers, were intentionally concreted during the last stage of construction. As these carry the self-weight of the completed parts, so pretensioning is achieved in order to avoid cracking of concrete during service.

The reinforcement was firstly placed on formworks, prepared to receive the concrete, and the complete arches and decking at the ends were poured. The middle third section of the deck is left to be poured in the following stage. Then, the scaffolding was removed. Thus, the hangers, tie beam and the middle deck were loaded with the self-weight of the structure. Finally, concreting of these remaining elements was completed. In this way, pretension was applied to these reinforcements. The second-stage concreting was done after 1 month.

Gediz bridge opening was announced in the newspapers and had been planned to be opened within 4 months[41] by Prime Minister İsmet İnönü.[42] However, the opening was instead attended by Governor and Nafia managers. The bridge was described in the title of the newspaper as "Republican and National Bridge".[43] Later news announced the bridge as an important public work.[44] Some of these articles are shown in Figure 4.40.

As a result of boring tests, it was determined that the ground consisted of thin and liquid mud for up to 25 m followed by clay. The foundations of the bridge were, therefore, rested on wooden piles. Piles were driven with difficulty and reached the hard suitable ground with a minimum socket length of 6–8 m. The foundations

Figure 4.39 Menemen bridge (author).

Gediz Köprüsü Dört Ay Sonra Açılacaktır.

Yeni Gediz Köprüsü İsmet İnönünün Eliyle Açılacak

(Cumurluk Büyük Gediz Köprüsü) Dün Törenle Açıldı

Gediz köprüsü önemli bir bayındırlık eseridir

Açılış töreni nasıl oldu?

Figure 4.40 Various newspaper articles for the Gediz bridge from top-left to bottom-right
  a. 'Gediz Bridge will be opened in 4 months' (21 February 1935), b. 'New Gediz bridge will be opened by [Prime Minister] İsmet İnönü' (13 June 1935), c. 'Big Gediz Bridge of Republic opened with a ceremony yesterday' (1 August 1935) d. 'Gediz bridge is an important public work' (4 August 1935).

were protected against scour with riprap, that is large graded stone placed in a random fashion around the piers or on banks of the river (Table 4.2).[45]

An opening ceremony of the bridge was held on 01 August 1935. Preparations were made, and a triumphal arch was erected. The ceremony was attended by the Governor of İzmir, General Kazım Dirik, public officials from the surrounding districts and a crowd of the local community. The first speech was made by the Chief Engineer of Nafia followed by the Mayor of Menemen. Later, Ali Çetinkaya closed the round to speeches and opened the bridge by cutting the ribbon.

Table 4.2 Menemen Bridge Design Information Summarized Regarding Foundations (author)

	Abutment Menemen Side	First Pier	Second Pier	Third Pier	Fourth Pier	Abutment Bergama Side
Excavation method	Timber sheet piling	Timber+iron sheet piling	Timber+iron sheet piling	Caisson	Caisson	Caisson
Depth from lowest water level	–2.7	–3.8	–3.2	–3.5	–3.88	–2.9

After the ceremony, all the delegation went over the junction road with cars then went to the wooden bridge and crossed over it for the last time in ceremonial jest.

BWS-12: Karabekir bridge

This bridge was built to connect two nearby towns, Çorum and Sungurlu, to the Ankara-Kayseri Railway line. The bridge is on the Delice river which joins Kızılırmak. This location was chosen for the configuration of the bridge, as the Delice bed becomes regular and narrows. Thus, it was possible to cross the river with a superstructure span of 32 m.

The total width of the bridge is 4 m, including 3 m of road section and 50-cm sidewalk. The rise-to-span ratio of the bridge is 1:5, which gives an arch height of 6.4 m. The cross section at the crown is 50 × 90 cm. Pretensioning in the same manner as Menderes (Gediz) bridge was applied to this bridge.

The project was awarded to the contractor Muhtar Bey with the contract dated 2 October 1934 and completed on 6 July 1935 (Figure 4.41).

BWS-13: Dalaman (Atatürk) bridge

The road connecting Köyceğiz and Fethiye to Muğla was a very important part of the network in the region, and a bridge was needed at Dalaman Stream, which is also known as Kocaçay (Figure 4.42).

A bridge for this crossing was initially planned by Abbas Hilmi who was a Governor in Egypt, an Ottoman land at that time. He ordered a bridge for accessing his farm in Dalaman. The substructure of the bridge was built, and the superstructure was imported from France; however, the bridge did not proceed because of World War I. The series of events for this bridge can be traced from records. A letter[46] was found in Ottoman archives, requesting the drawings of the superstructure so that its assembly can be resumed. The other is the bridge inventory found in KGM achieve,[47] which was probably the reply to this request for the 70-m span bridge. Lastly, the 1938-dated newspaper advertisement was found for the sale of a bridge superstructure as scrap and described it as "partly covered under fill". We can conclude that the superstructure was not assembled at all.[48]

Figure 4.41 **Karabekir bridge under restoration (author).**

Figure 4.42 **Dalaman bridge (FATEV archive).**

This original substructure has since been left abandoned as its location is prone to floodwaters (Figure 4.43).

A Nafia report (1933) stated that trunks of trees and debris exceeding 20 m in length were carried during previous floods; therefore, it was necessary to make spans at least 35 m long. The total length of the bridge is 105 m and is made by three spans. The total width of the bridge is 4 m with 3 m road width (Figure 4.44).

DEMİR KÖPRÜ AKSAMI SATILIYOR

Bankamız malı olup Dalaman çiftliğinde kısmen toprak altında mevcud demir köprü aksamı pazarlık suretile satılacaktır. İsteklilerin 7 temmuz 938 perşembe günü saat 11 e kadar tekliflerini Bahçekapıda Taş handa 39 numaraya bildirmeleri, görmek istiyenlerin de Dalaman Çiftlik Müdürlüğüne müracaatleri.

Figure 4.43 "Iron Bridge for Sale!" Cumhuriyet newspaper archive dated 18 April 1938.

Figure 4.44 Dalaman bridge design drawings (KGM archive).

Dalaman bridge was tendered to SAFERHA with a contract signed on 14 October 1934. The construction of the bridge was completed on 8 June 1936 (Figure 4.45).

Today, however, the road network has been modified, and the bridge has been bypassed by the main road.

BWS-14: Yalakdere bridge

This structure was built on the Yalova-Kocaeli road. Currently, the bridge is no longer part of the main road, and it is parallel to the main arterial (Figure 4.46).

Although there appear to be no records of any maintenance works undertaken, the structure is still in relatively good condition. The parapets are metal railings with concrete posts at pier locations and appear unmodified since originally built (Figure 4.47).

The bridge was initially designed with two beam spans each of them 10 m long with a masonry pier in the middle of the stream. Unfortunately, the bridge failed with its pier scoured and partly collapsed still carrying the beams. Subsequently, the crossing was designed with a longer-span bridge and an increased elevation.

The two-span bridge was converted to three-span bridge with a central bowstring span of 22.50 m. It has 10-m approach spans on both approaches, and the total length of the bridge is 53 m (Figure 4.48).

The side spans of the approaches of the current bridge were reused from the partly failed bridge. The superstructures, weighing 80 tonnes per span, were first pulled to

Figure 4.45 News with title "Dalaman bridge construction is very advanced stage" from Anadolu newspaper (26 May 1936).

Figure 4.46 Yalakdere bridge (FATEV archive).

the sides and placed on the embankments of the bridge approaches, then placed at their final position following the construction of the new piers. The bridge road width is 4.80 m. The pier foundations rest on wooden piles.[49]

Figure 4.47 **Yalakdere bridge (author).**

Figure 4.48 **Yalakdere bridge (author).**

During the author's visit in 2015, the water level was very low, and the collapsed pier was visible in the river bed below the bowstring span.

Yalakdere bridge was contracted on 4 November 1935 to SAFERHA, and the construction was completed on 10 December 1936 (Figure 4.49).

Figure 4.49 Yalakdere bridge (author).

Whilst reading about the Yalakdere bridge from Akkaya, we also learnt about the transportation networks which were very poorly interconnected at that time. He had to take a train, boat, or car, and undertake fair amount of walking to finally reach the bridge site, which is in fact approximately 80 km away from İstanbul.

Akkaya[50] summarizes his visit to this bridge with the following description:

> I was reaching the Yalakdere Bridge, by train to Gebze and on foot to Eskihisar, then by crossing in a boat to the other side. It was back then a forbidden area between Pendik and İzmit. It was forbidden to take a photo or get off the train without permission. A police officer must be met at the station for querying such permission. There was no highway between İstanbul and İzmit.
>
> We were restoring and upgrading an old bridge that collapsed during the flood in Yalakdere. The construction works and management of this bridge, which is difficult to access, was very challenging. İzmit Nafia appointed a project controller to the works being undertaken for this bridge; he was a tall and slim technical officer. This young man was so neglectful to the work and the world, reading left wing [political] publications with awe as if reading the Quran...
>
> One day, we heard that the huge subcontractor pushed the surveyor into the foundation pit being recently excavated, with wooden piles driven and concrete poured. Nafia stopped the job! It was an issue to travel to the bridge from İzmit with the Manager of Nafia and recommence the construction works by providing peace between the surveyor with a suspended arm and the huge subcontractor.

BWS-15: Sakarya bridge

As described in a 1937 newspaper[51], "Bridge is on one of Turkey's busiest roads with 1200 vehicles per day on the Adapazarı-Bolu road for the replacement of the wooden bridge.

The traffic density of this road became a big problem when the old bridge created a bottleneck. Nafia decided to replace the old bridge with a concrete one and tendered the project in 1935" (Figure 4.50).[52]

The bridge is crossing over the Sakarya river with three spans of 35 m. During the visit of the author, it was possible to identify the expansion joint over the pier which continues at the railing as well. The width is 5.60 m between the arches. The arches are 60 cm wide and are 100 and 125 cm in depth at the crown and springing, respectively (Table 4.3).

The upper layers of the ground are sand followed by soft clay mud. The piers are surrounded by riprap protection against scour especially during the flood. Even though

1. Bl./16 KÖPRÜ ADI : Sakarya
ILI : Sakarya
YOLU : Adapazarı - Akyazı Kav.
K.K. NO. : 54-02 KM. : 3+400
ÜZERİNDE BULUNDUĞU SU ADI : Sakarya

CİNSİ - TİPİ : Betonarme Bowstrink
TÜM UZUNLUĞU : 108.60 M. GENİŞLİĞİ : 70+480+70 SM.
EN BÜYÜK AÇIKLIĞI : 35.00 M. (3 Açıklıkiı)
PROJE HESAP YÜKÜ : H — S
HİZMETE GİRDİĞİ TARİH : 1937

51

Figure 4.50 Sakarya bridge (KGM album 1973).

Table 4.3 Sakarya Bridge Total Amount of Materials Used (author)

	Superstructure	Substructure
Larssen sheet pile L=7m	-	50 tonne
Reinforcement in concrete	102 tonne	
Concrete	1243 m^3	
Timber pile, 30 cm Dia. L=10m	-	128 each/101 m^3
Total weight of structure	1416 tonne	1836 tonne
Riprap – bridge piers		500 m^3
Riprap – riverbanks		550 m^3

6 m of scour depth was observed during the construction of the Adapazarı side pier, no scour was observed after the riprap protection was applied.

During the site selection of the Sakarya bridge, it was observed that the flow eroding the existing timber bridge approaches on the Adapazarı side. Therefore, a more suitable site was sought for this particular bridge; however, it was understood that there were no other stable and firm ground available and any alternative locations resulted in the bridge becoming unfeasibly longer. Therefore, it was decided to construct the new bridge at the location of the old wooden structure and make protection on riverbanks. In addition, the stream flow would be modified to flow in a more perpendicular direction to the bridge axis to reduce the risk of scour (Figure 4.51).

At the same time, the riverbanks were proposed to be protected with wooden breakwaters or tree branches; however, this arrangement was considered to be insufficient, and riprap was chosen. Suitable stone for this task was not available in the Adapazarı region. It was necessary to transport the stones from Doğançay to Sapanca by train. Then, piers of the bridge and 270 m of the riverbank on upstream side of the

1. Bl./16	KÖPRÜ ADI	: Sakarya	CİNSİ - TIPI	: Betonarme Bowstrink
	İLİ	: Sakarya	TÜM UZUNLUĞU	: 108.60 M. GENİŞLİĞİ : 70 + 480 + 70 SM.
	YOLU	: Adapazarı - Akyazı Kav.	EN BÜYÜK AÇIKLIĞI	: 35.00 M. (3 Açıklıklı)
	K.K. NO.	: 54-02 KM. : 3 + 400	PROJE HESAP YÜKÜ	: H — S
	ÜZERİNDE BULUNDUĞU SU ADI	: Sakarya	HİZMETE GİRDİĞİ TARİH	: 1937

Figure 4.51 Sakarya bridge (KGM album 1973).

Figure 4.52 Sakarya bridge, the author's son pointing at the "broken" part on the deck (author).

bridge were protected. This stretch was the concave side of the bank, to which erosion and flow velocity are the greatest threat (Figure 4.52).

Once the bank protection was completed and it was proved to provide efficient protection, it was decided to preserve the 280 m adjacent to the bridge in the same manner. Usually, the riverbank protection was made in the form of rounded stones, but for the first time, stones were thrown as they were, that is irregular angular shape, expecting that such stones will settle with time and interlock. Since a contractor could not be found, this work was made "emaneten".[53]

The Sakarya bridge was tendered to SAFERHA with the contract dated 20 July 1935 and was completed on 31 May 1937.

Ceremony: All the prominent Nafia leaders, including the Nafia Deputy and the deputies and undersecretaries of the region, attended the opening ceremony. The train arrived at Arifiye station on the morning of 20 June 1937; İzmit Governor, Party Leaders, the Mayor and Mürsel Pasha were welcomed by other notable personnel. After Arifiye, the train reached Adapazarı station at 09:45 AM, and all were welcomed with a military ceremony.

Ali Talip (Güran) Bey, the Head of Highways and Bridges, gave information about the bridge at the ceremony:

"…For foundations, after the sheet piles were driven, the water was drained continuously and the foundation was excavated from the low water level to a depth of 3–4m and after that 30cm diameter and 10m long wooden piles were driven. The total amount of piles driven was 128 and foundation concrete was poured on top of them. The part of the foundation in the soil is 3–4m deep. In the meantime, the metal sheet piles around the foundation reached 4m deeper than the foundation bottom level, to protect pier from scour and undermining, the riprap protection was used around the piers."

Opening of the bridge was announced in newspaper[54] as a second happiest day of Sakarya river: 1921 Battle of Sakarya and 1937 Sakarya Bridge Opening.

BWS-16: Çetinkaya (Bafra) bridge

Köprüden geçti gelin, başbağın düştü gelin…
A well-known folk song starts, "Bride crossed over the bridge, hairband has fallen; you have passed me bride, but I cannot give up on you".
This folk song, Köprüden Geçti Gelin, narrates the story of a bride crossing the original wooden bridge at this location. The horses were suddenly frightened by a flying eagle, who had snatched the hairband of the bride which got the horses worked up, so they drove in the river and the bride and her entourage lost their lives.
In Turkey, there is a tradition that a bride and her entourage should cross a bridge, stand in the middle and throw an apple into the water. This tradition still survives in many places, and the famous one is the Çetinkaya bridge, in Bafra (Figure 4.53).
The Çetinkaya bridge, which is one of the well-known bridges in the country, is in the town of Bafra crossing over Kızılırmak (Halys) on the Samsun-Bafra-Alaçam road. It has seven bowstring spans of 35 m each.
Çetinkaya bridge was built in 1937 to replace the existing timber bridge at this location.
Nafia (1933) described the bridge as follows:

This bridge is very valuable because it is located on the main arterial road of the Bafra region, which has great importance in the country's economy with its

Figure 4.53 **Bridge photograph on the cover of from Bayındırlık Ministry Magazine (Year: 1937/7).**

tobacco production. It is among the most important of the road bridges built so far, as it is crossing over one of the largest rivers in Anatolia the Kızılırmak where it spills into the sea...

After the wooden bridge in the same location became worn out, a reinforced concrete bridge was needed. As a result of the on-site inspections conducted jointly with the Water Works Department, it was decided to build the new bridge 10 m away to the south and downstream of the former wooden bridge.

The spring floodwaters scoured one side of the riverbank and filled the other side with debris. Kızılırmak river, which constantly changes its shape due to the deposits brought by the flow, eroded both riverbanks, especially one side bringing a danger to the town of Bafra in 1937.

Although the river bed width was around 600 m at the old bridge site, it increased up to 1 km on both the upstream and downstream sides. In order to cross this river with a bridge of 255 m length, nearly half the length of the course, it was necessary to manage the flow, and substantial riverbank protection and embankment construction were considered.

At the same time, during the investigation studies of 3 years, which involved regular measurements for flood levels, flow speed, etc., it has been observed that the west side of the riverbank was in the process of eroding.[55] Therefore, the plan was to redirect the water flowing from the west side, which was separated 450 m at the upstream of the bridge. This was attempted by driving a wooden sheet pile and throwing leather sandbags in front of this river branch. However, due to the depth and the natural slope of the river bed, which was aligned to the west side, and also the contributing seasonal floods, this attempt was not successful.

Measures taken to manage the river flow during the construction works are shown in a detailed plan sketch, in the Nafia Magazine.[56] In this sketch, it is interesting that the tree branches of 4–5 m were used, as described in 1932-dated paper of Hayırlıoğlu, together with gravel in layers to form an embankment to protect the riverbank from the flow (Figure 4.54).

In the concept stage, the small span lengths were avoided to decrease the number of foundations, which needed to be too deep to be protected against scour. The foundations were sunk up to 4 m below the lowest water level and rested on wooden piles. The cofferdam was made with 8.5-m-long Larssen sheets, which are a type of sheet piling, named after its developer Tryggve Larssen (Figure 4.55).

The length of the bridge is 255 m, and the width is 6.8 m. The tender was awarded to Mahmut Hüseyin, Mustafa Reşit and Partners Collective Company on 30 July 1935 and completed on 15 October 1937.

In addition, since the river was diverted during the construction of the bridge, the flow had to be turned back under the bridge again. Bafra Lumber Factory took responsibility for this task.[57]

The opening ceremony of the bridge was held on 4 November 1937. An enormous crowd and guests from Samsun and Alaçam attended the ceremony.

At the request of the people of Bafra, the bridge was named "Çetinkaya" to honour Ali Çetinkaya.

Nowadays, a new bridge of the same length, with spans of 17.5 m (half of Çetinkaya), has been built next to the bowstring bridge.

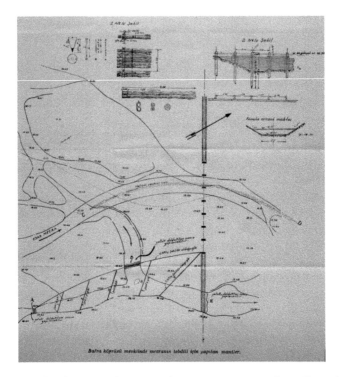

Figure 4.54 Çetinkaya bridge river diversion during construction from Bayındırlık Ministry Magazine (Year: 1937/11).

Figure 4.55 Pretensioning of Çetinkaya bridge hangers before concreting (SALT research, photograph and postcard archive).

BWS-17: Ergene bridge

Ergene bridge was on the Muratlı-Büyükkarıştıran road and crossing over the Ergene river. This local road connects with the İstanbul-Edirne main road (Figure 4.56).

The preliminary projects were prepared by the Bridge Design Team of Roads and Bridges Directorate within Nafia. The road profile was determined, and road projects were carried out by Nafia engineer Yusuf Onur. Bridge projects were prepared by the contractor.

The tender was awarded to SAFERHA on 17 January 1936. The contract was signed on 6 February 1936, and the works were formally started upon delivering an advance payment.

After being awarded the project, the contractor first established the construction site by finishing the preparations such as warehouse, storage and service building for the construction material.

The bridge foundations were all built over reinforced concrete piles for the abutments and three piers. Boreholes were not carried out during the design phase; therefore, a clear idea about the ground could not have been certain. The piling work started first by driving a test pile to determine the refusal values.

Initially, the work progressed according to schedule, and the foundations, substructure and deck section were completed, due to the contractor's late supply of timber for the scaffold, arch construction was delayed for about a month, and when the winter season started concreting had to be put on hold, this resulted in about 71-day delay to the scheduled construction programme.

During this period, the contractor worked under a penalty as a result of delays, and at the end of this period, he applied for temporary acceptance. It was decided that the acceptance procedures for the handing over of the bridge should be done by the local Nafia branch. Tekirdağ Nafia manager, Adli Pelin, and engineer Sermed from the Bridge Design Team in the main office and the project control engineer Yusuf Onur carried out

Figure 4.56 **Ergene bridge (author's collection).**

Figure 4.57 **Ergene bridge (FATEV archive).**

the handover. The temporary acceptance letter was prepared by undertaking the necessary measurement and loading tests of the finished structure on 6–7 May 1938.

During the construction, a photograph album illustrating each construction stage
phase was prepared in accordance with the specifications and submitted to the Directorate of Roads and Bridges (Figure 4.57).

This bridge has three approach spans of beam type and the main bowstring span
of 31.33 m. The bridge was completed on 15 February 1937.

A possible location of the bridge is given in the list; however, the structure was not
found. The bridge was made redundant by road scheme changes and presumably did
not survive today.

BWS-18: Miliç bridge

According to KGM records, this bridge was built in 1939 in the district of Terme
in Samsun on the Ordu-Samsun road. The bridge is in the town of Miliç, and although it is now outside the main road network, it is open and serves local traffic
(Figure 4.58).

The contractor of the bridge is given as Hasan Selahattin Berkeman[58] in the KGM
record (Figure 4.59).

Figure 4.58 Miliç bridge (author).

BWS-19: Gülüşkür bridge

Gülüşkür bridge is a five-span bridge submerged in the Keban dam lake since the 1970s.

The Gülüşkür bridge spans were 36m long each. It was built in 1939 on the road to Elazığ to Palu on the Murat river, a branch of the Fırat (Euphrates).

The bridge was constructed by a renowned company called Aral Construction which also participated in the construction of Singeç (ACH-20) and Pertek (ACH-27) bridges in the region.

Aral Company was founded by Ragıp Devres, who was a student in the Engineering School when the First World War commenced and he went to the Caucasian front. The Caucasian front was especially challenging with snowy days, frozen nights and terrible diseases. He survived typhus and miraculously returned home alive as one of the only 270 survivors out of 15,000 strong corps.

Ragıp resumed his education and graduated from Engineering School in 1922. He then started his career as Bursa Province engineer but soon realized that office work was not for him and soon resigned. He then worked locally on the reconstruction of assets including some stations on the Anatolian-Baghdad Railway line which were destroyed during the war.

Ragıp Devres brought the first mechanical grader to Turkey to use in Kayseri-Sivas Railway line in 1929. He is recognized as the first engineer in Turkey to use this kind of modern machinery in his work (Figure 4.60).

Figure 4.59 Miliç bridge (author).

Figure 4.60 Gülüşkür bridge during opening ceremony (SALT research, photograph and postcard archive).

Figure 4.61   Horasan bridge during its construction from Bayındırlık Ministry Magazine (Year: 1939/5).

BWS-20: Horasan bridge

The bridge[59] is crossing over the Aras river on the Horasan-Ağrı road at the eastern entrance of the Horasan district. It was built in 1940.

It has three spans with a bowstring type of 36 m each. The total length is 110.8 m, and the bridge width is 5.5 m in total, with 4.5-m road section and 50-cm sidewalk on both sides. This bridge has six K-shaped transverse bracing members interconnecting the main arches (Figure 4.61).

The contractor of this bridge was given as Kemal Gençspor[60] in KGM records.

BWS-21: Kavuncu bridge

This bridge is located over the Sakarya river and is about 3 km east of Kavuncu village on the Günyüzü-Polatlı highway. The bowstring bridge[61] with a span of 30 m was built in 1945.

As the previous bridge of the same name was destroyed during the independence struggle, the current bridge was built as a replacement.

BWS-22: Beydağ bridge

Beydağ bridge is located in the Beydağ district of İzmir, on the road of Ödemiş-Beydağ. It crosses the Küçük Menderes river. The bridge has a span of 35 m and is currently outside of the road network.

BWS-23: Melen bridge

Melen bridge is currently on a local road that is parallel to the main state road of Adapazarı-Bolu, crossing over the Melen river (Figure 4.62).

Figure 4.62 Melen bridge (author).

Figure 4.63 Nif bridge (Photograph: Efkan Sinan/Flickr).

This bridge was built by STFA. Akkaya briefly mentions that, "they (STFA) had committed to the construction of the Melen bridge on Düzce road".

The span is 42 m, and it is very similar to the Nif bridge with the same arch detail. The entrance to the bridge still has the original railing details.

STFA records indicate the date of the construction of this bridge as 1945, whereas in KGM records, it was built in 1948.

BWS-24: Nif bridge

No records could be found about the bridge other than the heritage registration and tender announcement of its construction. Today, it is bypassed by the road network of Manisa-Turgutlu road, in the proximity to Turgutlu (Figure 4.63).

The tender announcement of the bridge was announced for 15 July 1948 in the Cumhuriyet newspaper. Based on this, we can conclude that the bridge was built between 1948 and 1950.

The bridge is currently registered by the cultural heritage conservation and described in its report as follows[62]:

> Nowadays, the bridge is quite worn and neglected. With the effect of precipitation, colour changes, cracks and ruptures have propagated, especially along the arches. In some places, reinforcement is exposed. Some of the hangers transferring load from the arch have been replaced with steel elements. It also appears that the parapets between the hanger have been modified.

NOTES

1. Beamish, R. (1862) *Memoir of the Life of Sir Marc Isambard Brunel*. Paperback. London.
2. Sutherland, J. (2009) The Birth of Prestressing Iron Bridges for Railways 1830 to 1850. *The International Journal for the History of Engineering & Technology*, 79:1, 113–130. DOI: 10.1179/175812009X407213.
3. Verantius, F. (1595) *Machinae Novae Fausti Verantii siceni Cvm Declaratione Latina, Italica, Hispanica, Gallica Et Germanica*. Venetiis. Cum Privilegiis.
4. Fulton, R. (1796) *A Treatise on the Improvement of Canal Navigation*. I. & J. Taylor at the Architectural Library. London. Page: 121. Plate 14.
5. Jordan, J. (1797) Patent for constructing aqueducts and bridges, with curved ribs of timber, or iron, and suspending therefrom, by iron rods, the floor or trunk, as the case may be. *London Repertory of Arts*, Vol:6. Page:220.
6. Griggs, F. G. (2014) *Theodore Burr's Trenton Bridge*. American Society of Civil Engineers. USA.
7. Smiles, S. (editor) (1883) *James Nasmyth, Engineer: An Autobiography*. Harper Brothers. New York.
8. Holzer, S. M. and Knobling, c. (2020) The laminated arch in the first half of the 19th century: A status report from Switzerland. Retrieved from: https://www.arct.cam.ac.uk/.
9. The Application of Spiral Hooping to a French Concrete Bridge (1909) Engineering News. Vol:61 No:16 Page: 438. New York. Retrieved from https://archive.org/.
10. A Criticism of the Reinforced-Concrete Bridge Truss (1909) Engineering News. Vol:61 No:16 Page: 440. New York. Retrieved from https://archive.org/.
11. Marsh, C. F. (1904) *Reinforced Concrete*. D. Van Nostrand Company. New York.
12. Creighton, W. F. (1909) Concrete Work on Sparkman St. Bridge, Nashville, Tenn. Engineering News. Vol: 61 no: 8 Page: 199–220. New York. Retrieved from https://archive.org/.
13. Concrete Truss Bridge: The First in Canada (1909) Canadian Engineer. No. 20. Canada. Page: 566
14. McFetrich, D. (2019) *An Encyclopaedia of British Bridges*. Pen and Sword Books. UK. ISBN 978-1-52675-295-6.
15. Chrimes, M. (1997) *Historic Concrete: Background to Appraisal*. Proceedings of the Institution of Civil Engineers - Structures and Buildings. The development of concrete bridges in the British Isles prior to 1940. Volume 122. Issue 4.
16. The Application of Spiral Hooping to a French Concrete Bridge (1909) Engineering News. Vol:61 No:16 Page: 438. New York. Retrieved from https://archive.org/.
17. BOA PLK-p-02082-0001 Bursa-Atranos yolunda Kocasu Nehri üzerine yapılacak betonarme köprü projesi ve resmi. (Fr.)
18. Nettleton, D. A. (1977) *Arch Bridges*. Bridge Division Office of Engineering Federal Highway Administration U.S. Department of Transportation. Washington, DC.
19. The term used for Public Works was "imar", means civilization and development excluding the social content. Term later changed to "Nafia", which can be translated as utility, to refer

to public assembly at the governance level. Nafia also later changed to "Bayındırlık", means prosperous, in 1935.

20. Nafia (1933) *On Senede Türkiye Nafiası 1923–1933*. Nafia Vekaleti Neşriyatı. İstanbul.
21. University was established by Jean-Rodolphe Perronet, and has raised many respected names in the field of bridge engineering.
22. Cumhuriyetin İlk Köprüsü (2017) Retrieved from: http://odemisyazilari.com /index.php/2017/10/03/cumhuriyetin-ilk-betonarme-koprusu-adagide-koprusu/.
23. Gazi means War Veteran. Name used as Gazi Mustafa Kemal until 1934 Atatürk surname have given to nations leader by parliament decision.
24. 1929 Yılında Yapıldı... Ankara Köprüsü... (2017 December 12) Pusula Newspaper. Retrieved from: http://www.pusulagazetesi.com.tr/arsiv_90409/1929-yilinda-yapildi-ankara-koprusu/
25. The Silahtarağa Power Station (Turkish: Silahtarağa Elektrik Santralı) was a coal-fired generating station located in İstanbul Turkey. The Ottoman Empire's first power plant, it was in use from 1914 to 1983.
26. Türkiye Mühendislik Haberleri (1969) Engineer Portfolios. Chamber of Civil Engineers Magazine. Ankara.
27. In 2 June 1929, a new law for "Road and Bridges" was in force. With the new law, the practice regarding the unification of state and provincial roads was abandoned. Then the management of road construction as again resumed by Nafia instead of Local Government Authorities in the provinces.
28. Silahtarağa ve Sünnet Köprüleri Münakasası (1930 August 1) Cumhuriyet Newspaper. Page: 5.
    Memlekette Köprü Siyaseti (1931 December 8) Son Posta Newspaper. Page: 4.
29. Sünnet Köprüsü: İnşaat Bitti Kaldırımlar bir türlü bitmiyor! (1932 June 15) Akşam Newspaper. Page: 3.
30. Nafia Vekaletinden: Kargir Köprü Münakasası (1930 June 19), Cumhuriyet Newspaper. Page: 8.
31. Nafia Vekaletinden: Köprü İnşaat Münakasası (1931 August 20) Vakit Newspaper. Page: 8.
32. Figure H.8 Report from the control engineer on construction of Akçaağıl Bridge 29 08 1932. Source: Unclassified documents from the State National Archives-Republican Archives. KGM Fund, Binder no: 2281. Örmecioğlu, H. T. (2010). Technology, Engineering, and Modernity in Turkey: The Case of Road Bridges between 1850 and 1960. Ph.D. thesis submitted to Middle East Technical University (METU).
33. A tremie is a watertight pipe, usually of about 25 cm diameter, with a conical hopper at its upper end above the water level. It may have a loose plug or a valve at the bottom end. A tremie is used to pour concrete underwater in a way that avoids washout of cement from the mix due to turbulent water contact with the concrete while it is flowing. https://en.wikipedia.org/wiki/Tremie.
34. Dose is the amount of kg cement in m3 of concrete.
35. BCA 30-18-1-2_16_82_17_11_20_2019 1_39_01: PM-Antalya-Serik yolu üzerinde Aksu köprüsünün inşaası.
36. Memleket Haberleri (1933 April 21) Yeni Köprünün Açılma Resmi Yapıldı. Akşam Newspaper. Page: 6.
37. **Feyzi Akkaya** (1907–2004) a prominent bridge engineer of the early Republican era. Together with **Sezai Türkeş** founded STFA Company, which was a leading engineering company in Turkey and worldwide. Their inventions and novel applications have been a vital contribution to the engineering field in Turkey. See Chapter 3 for expanded biography.
38. Akkaya, F. (1989) *Ömrümüzün Kilometre Taşları: STFA'nın Hikayesi*. Bilimsel ve Teknik Yayınları Çeviri Vakfı. İstanbul. Page: 36–37.
39. **Kemal Hayırlıoğlu** (1891–1984) was Nafia Chief Engineer for 15 years between 1928 and 1943. He worked in various positions in Nafia until retired in 1955. See Chapter 3 for expanded biography.
40. Photograph can be seen in Figure 3.2 of Chapter 3.
41. Gediz Bridge will be opened in 4 months (1935 February 21) Anadolu Newspaper. Page: 1.
42. New Gediz bridge will be opened by İsmet İnönü (1935 June 13) Son Posta Newspaper. Page: 4.

43. Big Gediz Bridge of Republic opened with a ceremony yesterday (1935 August 1) Anadolu Newspaper. Page: 1.
44. Gediz bridge is an important work of Art (1935 August 4) Kurun Newspaper. Page: 5.
45. Ergani, C. (1934) Menemende Gediz Nehri Üzerinde Yapılan Betonarme Köprü. Türkiye Cumhuriyeti Nafıa Vekaleti Bayındırlık İşleri Dergisi Sayı.1 Sayfa.1 - 9 ss.
46. BOA: H-15-02-1335, Yer Bilgisi: 73–8: Belge Özeti: Köyceğiz'de Dalaman Çayı üzerinde hidivce yaptırılırken harp dolayısıyla yarım kalan köprünün yeni aksamının ne şekilde kurulacak yerine konulması gerektiğinin tesbiti için bu aksamın bir resminin çizilerek gönderilmesi gerektiği.
47. Figure 4.11 List of the steel elements remained. From the inventory list produced in 1930s for construction of 70 m span steel town lattice truss type Dalaman Bridge. Source: Unclassified document from the State National Archives-Republican Archives, KGM Fund, Binder no. 2203. Found in: Örmecioğlu (2010) Page: 119.
48. Cumhuriyet Newspaper (1938 April 18) Demir Köprü Aksamı Satılıyor. Adversitement.
49. Cumhuriyetin 10+3 Yılında Türkiye Bayındırlığı (1936) Bayındırlık İşleri Dergisi. Administrative Part. Month: October Year: 3 Sayı: 5 Page: 67.
50. Akkaya (1989) Page: 69.
51. Sakarya Köprüsü dün Büyük Merasimle Açıldı (1937 June 21) Cumhuriyet Newspaper. Page: 1.
52. Haykır, Y. (2011) Atatürk Dönemi Kara ve Demiryolu Çalışmaları. Doctorate thesis submitted to Fırat University. Page: 648
53. Emanet: abbreviated for "emaneten yapım usulu" (force account work method) means the system of carrying out a construction project by public authorities itself, instead of performing the work through a contractor.
54. Sakarya Bridge opened with a Ceremony Yesterday (1937 June 21) Posta Newspaper Page: 5.
55. Hayırlıoğlu, K. (1932) Bafra Köprüsü, Köprü Başları ve Temelleri. Türkiye Cumhuriyeti Nafıa Vekaleti Bayındırlık İşleri Dergisi (3) 3.1932, Pages: 10–18 Milli Kütüphane Yer No: 620.
56. Sketch (1937) Bayındırlık İşleri Dergisi. Administrative Part. Month: April Year: 3 Sayı: 11 Page: 148.
57. Çetinkaya Köprüsü Geliş ve Gidişe Açıldı (1937) Bayındırlık İşleri Dergisi. Administrative Part. Month: November Year: 4 Sayı: 6. Page: 113.
58. **Hasan Selahattin Berkeman**: Bridge engineer, worked as a contractor in the Samsun-Sivas line in 1926–1928, he constructed the reinforced concrete tunnel at the 194th km of the same line in 1932 and worked as a subcontractor in 26th section of the Fevzipaşa-Diyarbakır line in 1933–1936. Source: Work completion letter accessed at: www.delcamp.net.
59. Photographs of the bridge can be seen at: http://kopriyet.blogspot.com/2016/06/horasan-koprusu.html.
60. **Kemal (Veli) Gençspor** (1915–1990) was a graduate of İstanbul Forestry Faculty in 1938. He worked in public institutions for 5 years after his graduation and started his career in the private construction sector in İzmir in 1942.[60] He was a new graduate when the bridge was built.
61. Photos of the bridge can be seen at: http://kopriyet.blogspot.com/2016/10/kavuncu-koprusu.html.
62. Decree (2015) T.C. Kültür ve Turizm Bakanlığı. İzmir II Numaralı Kültür Varlıklarını Koruma Bölge Kurulu Koruma Kararı. Retrieved from: https://korumakurullari.ktb.gov.tr/.

Chapter 5

Concrete Arch Bridges

5.1 INTRODUCTION

The structural arch was an invention of human ingenuity long before modern engineering had been established. The first arch forms were corbelled, which indeed worked in a different principle to what we normally think of as a stone-built arch; they cantilever out progressively with each layer until the gap is closed at the top. Later arches, also called true arches, are formed by stone voussoirs, held in place by the friction between them and forming an arch shape. The deadweight of the arch is resisted by the lateral thrust at the springing. Therefore, the arch is sometimes defined as a structure that withstands the thrust at both ends.

Stone arches were formed by voussoirs laid on a curved formwork and locked at the crown by a keystone. The stones are in some cases connected by clamps which were usually formed by pouring molten lead into a preformed hole. As soon as the arch shape is self-sustained and the thrusts are properly resisted at the skewbacks, the bridge stands. However, as well described by Da Vinci "an arch is nothing other than a strength caused by two weaknesses",[1] so meaning that the arch bridge is only stable after it is completed. Therefore, the arch structure must be entirely supported during the whole construction process.

Types and Terminology: Arch bridges can be classified and grouped according to many parameters, including construction material, structural articulation and the position of the deck relative to the arch. The bridges included in this chapter are all made of concrete. The available information for their structural design was not always complete; therefore, the grouping will be done with known parameters. The following image indicates some of the common elements of a typical arch bridge structure (Figure 5.1).

The arch may be constructed either as a single structural element called a "barrel" or by separate multiple "ribs". Barrel is the name of the arch if it extends the full width of the bridge and rib defines individual narrower arches used together. Ribs are generally interconnected by struts or lateral bracings. Constructing the arch with ribs arose from the need to reduce the material needed for the structure.

The structural articulation is defined by Benaim[2] as "all the measures taken to hold it [bridge] firmly in position while allowing it to change in length and width and to rotate at supports". The means, i.e. bearings, to control the movements of

DOI: 10.1201/9781003175278-5

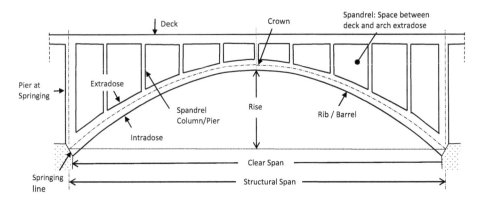

Figure 5.1 **Arch bridge terminology (author).**

the structure also transmit the loads to the foundations. These instruments can be fixed or hinged at the springing of the arch. At the same time, an additional joint can also be formed at the crown of the arch. Therefore, they are called "one-hinged" arch or "three-hinged" arch according to the total number of hinges in the arch structure. Even though KGM lists provide information on the arch articulation, this information is not complete for all the bridges and therefore will not be mentioned in this book unless relevant.

According to deck location; the bridge is called "through arch", if the springing starts at deck level and the arch curvature is above the deck. If the springing and the arch are completely below the deck, the bridge is called "deck arch". If the arch is positioned both below and above the deck, it is called "half-through arch"; in this case, springing is below the deck, but the arch extends above the deck.

The terminology is in general derived somewhat randomly. If "through arch" refers to the experience of driving through the structure, then the name "deck arch" is derived from the position of the deck relative to the arch. Indeed, a more straightforward and consistent terminology would be lower deck, upper deck and intermediate deck respectively, as also described by Troyano.[3]

One of the earliest definitions of "through Bridge" was made in 1908 as: "in cases where the floor system connects the bottoms of the trusses, the bridge is called a through bridge, as the traffic moves through the space between the trusses".[4] Indeed, this definition was initially made for truss bridges; later, the term became used in general to describe all types whose structural elements are above the deck.

A further definition for a through truss bridge implies the existence of cross-bracing above the deck. If bracing is not provided, it is called a pony truss, which normally has a shallow rise.

The geometry of the arch is another sub-classification. Early masonry arches, especially Roman arches were semi-circular, half of a full circle. Later in the Seljuk and Ottoman periods, the most commonly adopted form was the pointed arch, which has a pointed crown.

Public Works Authority/Nafia[5] favoured the segmental arch form, referring to it as "a result of modern engineering design calculations". Nafia's (1933) report[6]

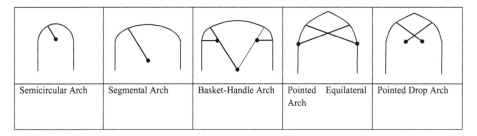

Figure 5.2 **Some arch geometry used in bridges in Turkey (author).**

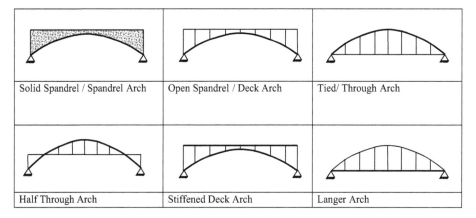

Figure 5.3 **Arch bridge types as referred to in the text (author).**

stated that old arch bridges of earlier eras were pointed or of semi-circular type, and that these arch types require more material and also limit the span lengths. Therefore, the segmental arch type was identified as preferable, as exemplified by the first Republican bridges, Garzan (BRD-15) and Agonya (Figure 5.2).

The first concrete arch bridges, which can also be considered as transition structures, were solid spandrel arches. In this form, the spandrel has fill over the arch up to deck level, also being called "spandrel arch". Later, the spandrel fill was replaced with spandrel columns or piers, and the structure became open-spandrel, which was then called "deck arch".

In closed-spandrels, the outside is walled in between the arch and deck; however, the enclosed space is hollow, as in the case of the Risorgimento bridge.[7]

Moreover, the solid/fill spandrel arch can also be called closed-spandrel, referring to the appearance of the bridge. Usually, it is not known from the outside if the spandrel volume is filled or not. In this book, the closed-spandrel refers to the arch where this volume is left empty (Figure 5.3).

Spandrel Arch: Solid-spandrel arch bridges are early types of concrete arch bridges imitating in appearance the earlier bridges. In solid type, the area between the travel surface (deck) and the arch ring is filled in, thus replicating the massive masonry bridge. The spandrel walls work as retaining walls, holding in the fill material, which could be earth, stones, rubble or some combination of materials.

The solid spandrel system has not been used for a long time, as engineers soon realized that significant material could be saved with a subsequent reduction of weight being achieved by eliminating the non-structural elements between the deck and the arch. Thus, open-spandrels were born despite the additional costs of constructing formwork for the spandrel columns.

Deck Arch: Open-spandrel concrete bridges evolved, as the span length of reinforced concrete arches increased and the weight and cost associated with the compacted fill, in conjunction with the consequent need for spandrel walls, became prohibitive. By eliminating these fill materials and walls, not only could dead loads be reduced but cost savings could also be achieved in structure. This also added lightness and openness with better visual appeal, especially for scenic locations.

Open-spandrel bridges, also called deck arch, transmit dead and live loads from the travel surface (deck) through spandrel columns or piers to the supporting arch. The arch ring may be either a barrel or ribs.

From the 1940s, the deck arch concrete structure began to be replaced by the more economic pre-stressed beam and reinforced concrete girder structures.

Through Arch: In this type, the springing of the arch starts at the deck level. As the arch is over the deck, the total rise is theoretically not constrained, therefore, allowing the achievement of much longer spans.

Half-through arch bridges have their springing under the deck level as an independent foundation. The arch, beginning below the deck, then extends above the deck level, allowing the effective use of vertical space for the bridge. In practice, the half-through bridge is a solution for shallow gorges which do not provide enough height for the arch rise.

5.2   HISTORY OF ARCH BRIDGES

The first arch bridges were made of stone with spandrel walls encasing the inside volume filled with rubble. In some cases, this fill material consolidates over time and becomes self-supported, even if part of the barrel collapses. However, most of the time, the fill in this volume exerts pressure on the spandrel facing walls, creating durability issues especially when proper drainage is not provided. This also explains why some stone bridges were constructed with spaces within spandrel walls to decrease the weight and also to eliminate the fill and its disadvantages.

The arch is by its nature more stable under load, provided that the load is uniform along the arch and symmetrical around the crown axis of the arch. In this way, the spandrel fill is advantageous for its weight spread over the arch as stabilizing effect and also distributing the exerted loads on the structure. However, the self-weight of the compacted fill is not uniform along the arch in consideration of the horizontal road surface, hence, creating a trapezoidal-shaped force with a resultant component acting on the arch closer to the springing. This uneven force can be regarded as another reason for the stone bridges being designed with reduced spandrel fill, as they evolved from being humpbacked to flat decks for wheeled traffic.

The transition from stone to concrete arch seems like a natural step up in terms of material. However, the same structural system with different materials can differ in many aspects. For example, masonry arch bridges could withstand significant deformations without loss of integrity due to their mass. The displacements for stone bridges are mainly a result of differential settlements of supports, whereas concrete arches are also exposed to deformations due to long sustained loading (creep), volume shortening (shrinkage) and thermal differences as a result of variations in weather conditions. Therefore, even though a mass concrete arch was suitable for short spans, reinforcement was necessary with the increase in span length.

Conventional masonry arch design had relied on theories like the middle third rule, which assumed the thrust to stay within the middle third of the section, so avoiding any tensile stresses within the arch. This condition is straightforward to satisfy for symmetrical loads; however, in the case of asymmetrical loads, like the one produced by a crossing vehicle, so-called live load, there is a moment caused along one side of the arch, with a sag on one side and hog on the opposite of the axis, therefore creating tension within the arch. Stone arches normally have a small proportion of live load in comparison to their permanent dead load on the bridge; therefore, the resulting combination of forces is not translated into net tensile forces. Concrete arches require different considerations, as they are required to carry a live load that is considerable in comparison to their own dead load. Therefore, the reinforcing of the concrete arch was again a structural necessity.

For both stone and concrete arch bridges, the same construction challenge of requiring temporary structures for their support is equally necessary, as concrete also does not gain its strength for some time, requiring full support during construction. The difference is in the process of construction: concrete is poured in formwork moulds; however, the stone is carefully assembled on top of the formwork, after being cut in the quarry and transported to the site.

The first reinforced concrete bridge ever recorded was built in 1875 by Monier[8] in Chazelet, central France. It was a 4 m wide pedestrian bridge, reaching a span of 16.5 m.[9] A bridge reminiscent of this that adopted similar parapets was built in Yıldız Palace, İstanbul, during landscaping works between 1894 and 1896 (Figure 5.4).

The earliest applications of concrete in arch bridges used a barrel arch filled up to meet the road level, resulting basically in a solid-spandrel bridge. These types of bridges generally have a reinforced arch; however, a few applications are seen using mass concrete, especially in countries that struggled to supply reinforcement.[10] One of these rare examples found in Turkey is the 1931 dated Agonya bridges, which appear to have been built in concrete without a single steel reinforcing bar.

In Turkey, the use of concrete in bridge structures first emerged as mass concrete used in foundation works. One of the earliest examples is found in the Ottoman archive in the drawings for an 1868 dated bridge project from Ayvalık-Bergama. This timber single-span superstructure was to be carried on concrete abutments.[11] It is not known if the project was implemented. However, the 1894 Daday–Kastamonu road project[12] known to be completed, has similar bridges, was also designed with foundations made of mass concrete.

For reinforced concrete, one of the earliest Turkish applications in infrastructure works was found to be undertaken in the Golden Horn port quay project in 1907. It is probable that it was constructed with imported cement, because the first cement factory, Aslan Cement, was established in Darıca in 1910 by a Danish investor.

Figure 5.4 Landscape bridge for the Yıldız Palace (TBMM Library).

Reinforced concrete was already used in many buildings in İstanbul and İzmir. Hennebique[13] had a local agent in both cities. In Turkey, Hennebique had one agency in 1904 and expanded dramatically to 28 licensees managed by an İstanbul (Constantinople) agency in 1913.[14] The İstanbul agent's name, André George, also appears in some drawings for Karanar Bridge crossing over Kızılırmak on Boyabat-Sinop road. It is not known, if the bridge construction was undertaken; however, the design drawings were found in archives.[15] The design shows a deck arch type with three ribs (Figure 5.5).

Revolutionary use of concrete was observed in Maillart's[16] innovative concrete bridges, that he designed together with Ritter.[17] He improved his concrete arch designs continuously from the field observations. In other words, his designs were the outcome of real-life experiments. In his bridges; the interconnected deck and arch, both contribute to load carrying capacity, like the flanges of a beam in which the arch itself and deck form the bottom and top flanges respectively. A well-known example is the Salginatobel bridge in Switzerland, which is a three-hinged arch with a 90 m span, finished in 1930.

His second innovation in design, called deck-stiffened arches, used a thin arch with a stiff deck. The 1933 dated Schwandbach Bridge in Switzerland with a span of 37.4 m is his best-known deck-stiffened arch bridge.

As Billington[18] states:

Meanwhile, in Zurich during the 1920s, with Wilhelm Ritter dead, the German scientific influence had increased, and Maillart's ideas began to be attacked by a new generation of academics, ...This attack came most heavily against Maillart's second new bridge form: the deck-stiffened arch...Even in Switzerland, researchers carried out major studies during this period and completely neglected Maillart's designs, not because they were unaware of them but because they were overwhelmed by the complexities of general theory.

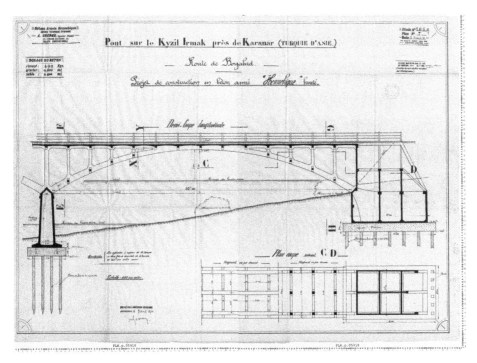

Figure 5.5 Karanar bridge project prepared by the Hennebique office in 1914 (BOA 2681, 4978, 5918).

This discrimination was also observed in Turkey for the Kömürhan bridge, in which the tender specifications did not permit the use of a deck-stiffened arch.

5.3 SIGNIFICANT FEATURES OF CONCRETE ARCH BRIDGES IN TURKEY

In the early Republican years, different arch types were used: like solid-spandrel, closed-spandrel, open-spandrel and arch with ribs. In terms of the arch itself, barrel types dominated the design with a few exceptions using ribs to form the arch. The collection of arch bridges had a variety of types; however, some types were only adopted once or twice. The majority of arch bridge types were the deck supported on spandrel columns which rest on a barrel arch (Table 5.1).

The first arch bridges, Irva and Kirazlık, with remarkable spans of 47 m, were barrel type with spandrel piers. The Nişankaya bridge also had a structural system with barrel arch and spandrel piers; it had two spans, each of them 18 m.

A unique collection in terms of type is the barrel arches with solid-spandrel type. Only two examples are known: Big Agonya and Small Agonya bridges, they used the same design with different span lengths.

Göksu (Beykoz) bridge with its single span is the only representative of a closed-spandrel type, in which the spandrel volume is not filled.

Table 5.1 Arch Bridge Groups According to Their Features (author)

Number of Bridges	Bridges Name	Arch Type	Spandrel Type
3	Irva, Kirazlık, Nişankaya	Barrel	Open spandrel with piers
1	Akçay (ACH-03)	Three Ribs	Open spandrel with columns
3	Big Agonya, Small Agonya	Barrel	Solid spandrel
1	Göksu (Beykoz)	Barrel	Closed spandrel
2 +29	Bakırçay, Fevzipaşa and All other bridges	Barrel	Open spandrel with columns
8	Aslan, Güreyman, Çayköy, Mayıslar, Dergalip, Sirya, Hasankeyf, Cizre (wide ribs)	Two ribs	Open spandrel with columns
6	Körkün, Meşebükü, Akçay (ACH-26), Km 386, Mutu, Sansa	Half-through	-
1	Balya	Four ribs	Open spandrel with columns

Bakırçay and Fevzipaşa, implemented using the same design with very slight modifications, were also constructed with barrel arch; however, they had spandrel columns instead of piers.

Another group of bridges, Aslan, Akçay and Güreyman, were the only bridges with ribs and spandrel columns supporting the deck. After the official establishment of KGM in 1950, arch ribs were only observed in seven bridges: Çayköy, Mayıslar, Dergalip, Balya, Sirya, Hasankeyf and Cizre. These structures mostly had two ribs and supported the deck with columns. Amongst them, Balya is a sole case with the provision of four ribs. These bridges have a pleasing appearance with their slender design and are unique representatives of their type.

After the official establishment of KGM as a new network authority in road infrastructure, bridge construction increased considerably. Standard designs were adopted especially for beam-type bridges. Typical solutions were used replicating existing designs and typical details, which were adapted according to local and specific conditions such as geotechnical characteristics at each specific site, required bridge width, road profile, etc. Arch bridges, however, were built based on independent and customized designs and differed in many details and features. Therefore, they are of significance in representing historical and engineering features.[19]

All the through-arch concrete bridges built during the early Republican era were tied arch type, in which the tie is used to withstand the thrust of the arch. This bridge type is also called bowstring, especially when applied to concrete. This bridge type is covered in detail in Chapter 4 of this book.

There are six half-through road bridges known to be constructed during the early Republican era. Almost all of them have a span range of over 60 m with a maximum span of 68 m for the Goat Bridge. An exception in this range was found in the Akçay bridge, where the span reached only 30 m.

These bridges are described in more detail later in this chapter in chronological order along with the other concrete arch bridges.

In the early years of the Public Works Authority, for the bridge designs carried out in-house, the span length range was 20–35 m for arch bridges. Exceptions to this were the 47 m spans for Irva and Kirazlık bridges: both structures were based on the same design prepared by a foreign company. Later, in 1932, a remarkable jump in terms of length was 109.6 m, achieved for the Kömürhan Bridge, which was constructed by NOHAB, who were constructing railway lines in Turkey. The next longest span bridge for the era was the 1939 Pertek Bridge with a 106.9 m span; this design was made by Emil Mörsch,[20] hired by a local contractor. After 1932, the average span range increased to 35–50 m, with the longest being 72 m in Körkün, 65 m in Meşebükü and 62 m in the Keban Madeni bridges. All of them were designed and built by Turkish firms. After 1950, the arch bridge span range was 50–70 m.

The relationship between the height and span of the arch, defined as the rise/span ratio, is a fundamental characteristic of an arch bridge, mainly for its structural design and also defines the appearance of the structure. The rise, in particular, and also the span may be determined through the conditions of the crossing. At the same time, the rise to span ratio is an initial and main parameter chosen for structural design and in return concludes the arch cross-section depth and the thrust to be resisted at end supports. Çayırhan and Maden bridges are two exceptions, as these were built with parabolic arches with a notably high rise/span ratio.

Construction and Special Applications: The construction of an arch bridge is one of the most challenging stages, even dominating the design since the arch is only stable once it reaches its finished state. During the entire construction process, it has to be fully supported on temporary structures. Various support methods are adopted during their construction. The simplest and most economical form was the wooden centring combined with formwork for concreting. Centring is a temporary structure built to provide the support and necessary shape for an arch during its construction. For longer spans, centring was supported as cantilevers from towers erected on piers and from the banksides. Nearly all bridges of the early Republic era were constructed with wooden centring, except the Malabadi bridge which used an arch made of steel as scaffolding after the previous centring was washed away by a sudden flood during construction.

Another method of scaffolding was the Melan system, which used a rigid steel reinforcement that could also act as scaffolding during the construction process. This alternative was considered during the tender process of the Kömürhan bridge but was abandoned as the amount of steel scaffolding reinforcement required for this system was found to be more than the required reinforcement of the arch, making it very expensive and unfeasible, especially in a country with limited steel supply.

Another innovative method is to construct half of the scaffolding arch on the bank and then rotate it to its designed place. This method was applied for the 62 m span Keban Madeni bridge, in which scaffold arch segments were prepared beforehand and assembled in a near-vertical position supported on the springing of the arch, then rotated inwards to its position for concreting. Figure 5.6 shows a similar method applied for a railway bridge from the same era, in which a scaffold arch built on temporary timber bridge, and rotated outwards by the auxiliary towers and cables.

Kömürhan bridge, the longest span built during the early Republican era with 109.6 m, was constructed on wooden scaffolding truss supported by provisional

Figure 5.6 Sivas-Erzurum line, km 374 railway bridge constructed using a similar method to Keban Bridge (STFA archive).

staying from wooden towers erected over the piers at the springing. Pertek bridge, the second-longest span, was constructed using a similar method.

Pasur bridge is a 50 m span, constructed with a wooden trussed-arch which acted as centring without any external auxiliary structure.

Half-through bridges with 64 m span in Körkün and 65 m span in Meşebükü were both constructed by SAFERHA. Körkün was also constructed on a self-standing wooden centring arch, incorporating the formwork for the structure. Whereas, the Meşebükü arch was constructed by provisional towers on each bank with stay cables supporting the arch. After the arch was finished, the deck was supported from the arch and the bridge was completed (Figure 5.7).

Among the other bridges, Alikaya bridge is known to have followed a different construction methodology in which an arch formwork was supported by fan-like shaped posts resting on the two temporary piers in the water, reducing the extent of the temporary foundations required in the deep riverbed.

5.4 EARLY REPUBLICAN CONCRETE ARCH BRIDGES IN TURKEY

Concrete arch bridges in early Republican Turkey were built from the first bridge in 1923, until today and will be a preference whenever conditions are suitable. Seventy bridges were introduced in this chapter as the significant structures for their heritage and engineering features from the period 1923 to 1967. This number covers all arch bridges built during 1923–1940 by Public Works, and beyond this date, only the bridges listed by KGM are included in the study (Figure 5.8).

Fifty-four bridges of the total 70 are individually described in the book; for the remaining bridges, there was not any additional information other than the records of KGM, and their current status and locations are provided in the lists given in Table 5.2.

Figure 5.7 **Körkün bridge hangers and deck under construction (İBB Atatürk Library).**

Among the total 70 bridges, 43 of them still stand in a good condition and most of them represententing the original features without any modifications to their structural system or appearance. Nineteen bridges out of 43, are still in use by road traffic, and the remaining 24 standing bridges are abandoned, mostly nearby or even just next to the new road network.

An exceptional case is the Km 441 bridge, which is a still operational railway bridge and is covered here because its design challenge is well documented and provides good insight into the construction management of the time.

Twelve bridges out of the total 70 are either already submerged in dam lake or will be submerged in the near future. For another group of 12 bridges, no information could be found for their current status, their presumed locations are provided in Table 5.2. Two bridges, namely Gençlik-1 and Gençlik-2 bridges, are known to be replaced. Gençlik bridges were utilized for pedestrian circulation within the park, their replacement seems very unfortunate since the heritage bridge was still in a fair condition to be used or to be rehabilitated. Notably, all the standing bridges are in good condition without any major maintenance work. As a single case, the Pisyar bridge has partially collapsed due to the flood.

The defining feature for an arch bridge would include the arch itself, spandrel columns or piers, or spandrel facing and springing details including the hinge supports. The detail for the continuation of the joint is of particular attention for these types of bridges. The crown hinge for Kirazlık and Kurtuluş bridges and the springing hinge for Irmak and Suçatı bridges can be mentioned as some to still preserve their original features for these details.

Parapets are also a characteristic of these bridges and some still retain their original significant parapets. Aslan, Big Agonya, Nişankaya, Fevzipaşa, Pasur (partly damaged), Güreyman, Meşebükü, Afrin, Fadlı, Akçay (ACH-26, partly remaining) and Suçatı bridges bear their original parapets mostly for the full length of the bridge.

Arch bridges were a very common application; therefore, the list provided here is not a full list. It covers many concrete arch bridges with a heritage potential starting

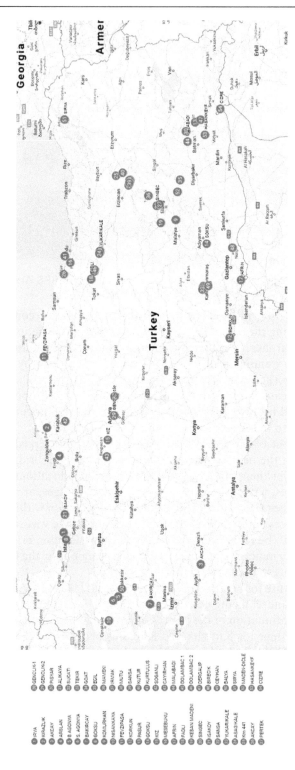

Figure 5.8 Concrete arch bridge locations shown on Google Maps (author).

Table 5.2 List of Early Republican Concrete Arch Road Bridges in Turkey (author)

ACH No	Name	Comp. Date	Location	River	No of Spans	Main Span Length (m)	Status	Contractor	Coordinates
1	Irva	1926	İstanbul	Irva	1	47	Incomplete information	İnşaat T.A.S. Societe...Urbaines	41.13564, 29.27472
2	Kirazlık	1928	Zonguldak	Kirazlık	5	47	Submerged	İsmail Hakkı	41.53342, 32.47481
3	Akçay	1928	Aydın	Akçay	2	35	Standing		37.80722, 28.31552
4	Aslan	1928	Zonguldak	Çaylıoğlu	2	24	Standing		41.23511, 31.61602
5	Büyük Agonya	1931	Balıkesir	B. Agonya	3	26	In use	Hayri Bey	39.83387, 27.33143
6	Küçük Agonya	1931	Balıkesir	K. Agonya	3	8	Incomplete information	Hayri Bey	-
7	Bakırçay	1931	İzmir	Bakırçay	2	26.76	Incomplete information	Saferha	39.05209, 27.09851
8	Göksu Beykoz	1932	İstanbul	Göksu	1	29	In use		41.08198, 29.06761
9	Kömürhan Ismetpaşa	1932	Malatya	Fırat	1	109.6	Submerged	Nohab	38.43931, 38.81574
10	Nişankaya	1932	Çanakkale	Sarıcaeli	2	18	Standing	Ankara İnşaat	40.05639, 26.59257
11	Fevzipaşa Ispiroglu	1933	Sinop	Gökırmak	2	26.76	Standing		41.60672, 34.64026
12	Körkün	1934	Adana	Körkün	1	72	Submerged	Saferha	37.09387, 35.21751
13	Pasur	1934	Siirt	Pasur	1	50	Standing	Saferha	37.96283, 41.79018
14	Göksu	1935	Adıyaman	Göksu	3	35	Standing	Salih Baran	37.692691, 38.082088
15	Güreyman Kız	1935	Ankara	Karabogaz	1	18	Standing		40.09042, 32.10528
16	Meşebükü Bolaman	1936	Ordu	Bolaman	1	65	In use	Saferha	40.98355, 37.50448
17	Afrin	1937	Kilis	Afrin	1	36	Standing	Fettah Aytaç Fenni İnşaat	36.80919, 36.98341
18	Fadlı	1937	Tokat	Fadlı	1	36	In use	Salih Baran	40.47846, 36.99704
19	Keban Madeni	1937	Elazig	Fırat	1	62	Incomplete information		38.80438, 38.72792
20	Singeç	1937	Tunceli	Hozat	1	36	Submerged	Aral	38.90244, 39.24274
21	Göksu Isakoy	1938	İstanbul	Göksu	1	29.1	Standing	Saferha	41.10767, 29.82877
22	Sansa KM 389	1939	Erzincan	Karasu	3	35	standing	Saferha	39.5538, 40.04721

(Continued)

Table 5.2 *(Continued)* List of Early Republican Concrete Arch Road Bridges in Turkey (author)

ACH No	Name	Comp. Date	Location	River	No of Spans	Main Span Length (m)	Status	Contractor	Coordinates
23	Yukarikale Koprübasi	1939	Sivas	Kelkit	1	43.5	In use	Muhtar Arbatlı	40.26728, 37.85788
24	Asağikale	1939	Sivas	Kelkit	1	36	Standing	Saferha	40.28718, 37.78465
25	Km 441	1939	Erzincan	Firat	1	45	Standing	Saferha	39.88978, 40.15987
26	Akçay	1939	Samsun	Akçay	2 (3)	30	Standing	Aral	41.13733, 37.15821
27	Pertek	1939	Elazığ	Murat	1	106.9	Submerged	Aral	38.83528, 39.2817
28	Gençlik Parki -1	1939	Ankara	Park	1	35	Replaced	Jack Acıman	39.9368, 32.84948
29	Gençlik Parki -2	1939	Ankara	Park	1	15	Replaced	Jack Acıman	39.936132, 32.851316
30	Pisyar	1940	Siirt	Garzan	1	45	Partial collapsed	Saha	38.18434, 41.52115
31	Alikaya	1940	Maraş	Ceyhan	1	45	Submerged	Saha	37.75456, 36.78989
32	Suçati	1940	Maraş	Ceyhan	1	35	Submerged	Saha	37.7628, 36.76244
33	Tekir	1940	Maraş	Tekir Creek	1	21.9	In use	Saha	37.78324, 36.69081
34	Km 386	1940	Erzincan	Karasu	1	68	Standing	Saferha	39.5665, 40.01794
	Keçi (Goat) Seyithan	1941	Tunceli	Hacisuyu	1	23	Standing	Salih Sabri Taşlıcalı	39.19245, 39.70543
	Dinar	1950	Tunceli		1	27	Incomplete information	Fikret Zeren - Ahmet Durak	
35	Dicle Eğil	1952	Diyarbakir	Dicle	1	80	Submerged		38.32916, 40.03632
36	Mameki	1952	Elazığ	Arapkir	1	72	In use	Saha	39.10121, 39.55379
	Gördes	1952	Manisa	Kumbaşı	2	42	Incomplete information	Ali Cumalı	38.9112, 28.29728
	Karasu	1953	Muş	Karasu	1	30	Incomplete information		38.63246, 41.92014
	Kuşcuderesi	1953	Elazığ	Kuşcu	1	36	Incomplete information		38.82017, 39.97537
37	Irmak	1954	Ankara	Kizilirmak	1	72	Standing	Vehbi Doğan	39.92362, 33.42723
	Keçihisar	1954	Kastamonu	Devrek	1	32	Standing		41.27086, 31.55561
	Mendo	1954	Bingöl	Mendo	1	34	Incomplete information		38.92428, 40.38038
38	Mutu	1954	Erzincan	Karasu	1	60	Standing	Saha	39.59006, 39.87243

(Continued)

Table 5.2 (Continued) List of Early Republican Concrete Arch Road Bridges in Turkey (author)

ACH No	Name	Comp. Date	Location	River	No of Spans	Main Span Length (m)	Status	Contractor	Coordinates
39	Sansa	1954	Erzincan	Karasu	1	80	In use	Saha	39.57442, 39.95177
40	Kütür	1954	Erzincan	Karasu	1	62	In use	Saha	39.73217, 40.2538
	Kizilmağara	1954	Erzincan	Karasu	1	61	Incomplete information	İzzet Odabacıoğlu	39.84176, 40.14935
41	Kurtuluş	1955	Ordu	Kurtuluş	1	20	In use		41.11294, 37.70993
42	Soğanli	1955	Karabük	Soğanli	1	68	Incomplete information		41.09987, 32.67296
	Sabuncu	1955	Kilis	Sabun	3	29.5	Incomplete information	Orhan Çobanoğlu	36.84318, 36.83852
	Karasu	1955	Bilecik	Karasu	1	32	Standing		40.1048, 29.99742
	Hamsu	1955	Bilecik	Hamsu	3	36.9	Standing		40.14763, 29.97667
43	Çayirhan	1955	Ankara	Sarıyar Dam	1	42	In use	Nasri Kaya	40.10448, 31.59559
44	Malabadi	1955	Diyarbakır	Batman	1	56.5	In use	G. Büyükyıldırım	38.15438, 41.20368
45	Çayköy	1955	Eskişehir	Sakarya	1	60	Standing		40.04448, 30.4545
	Mayislar	1955	Eskişehir	Sakarya	1	62	Standing	Bekir Bora	40.04262, 30.64567
	Dokuzdolambaç-1	1955	Tokat	Kelkit	1	60	In use		40.40814, 37.23428
46	Dokuzdolambaç-2	1955	Tokat	Kelkit	1	60	In use		40.39732, 37.29424
47	Dergalip	1956	Siirt	Botan	1	60	Submerged	Orhan Çobanoğlu	37.82338, 41.87385
	Güneysu	1956	Rize	Güneysu	1	70	Standing	Fikret Yüksel	40.98065, 40.61286
48	Birecik	1956	Urfa	Fırat	5	57	In use	Amaç Ticaret	37.02683, 37.97329
	Hasanlar Ballik	1957	Düzce		1	53	Standing		40.91365, 31.27104
49	Ceyhan Tabakhane	1958	Maraş	Ceyhan Tabakhane	1	60	Submerged	Ayhan Köker	37.63205, 36.80539
50	Balya	1959	Balıkesir		1	30	In use		39.74519, 27.56985
	Sultansuyu	1959	Malatya	Sultansuyu	1	37	In use	Nurhayr İnşaat	38.33887, 38.06357
51	Sirya Zeytinlik	1960	Artvin	Çoruh	1	67.5	Submerged	Sedat Baycan, Süleyman Betin, İzzet Bellikan	41.12055, 41.87292
52	Maden	1962	Elazığ	Dicle	1	40	In use		38.39532, 39.67651
53	Hasankeyf	1964	Siirt	Dicle	3	60	Submerged		37.71564, 41.41279
54	Cizre	1967	Cizre	Dicle	3	60	In use		37.33283, 42.19318

from 1923 to the last bridge dated 1967. The list of these bridges is given Table 5.2 with some information such as name, completion date, location, obstacle crossed (mainly river), number of spans, length of its main span, current status, contractor and coordinates.

ACH-0I: Irva bridge

The bridge is on the Beykoz-Bozhane road within the city borders of İstanbul. No information was found for this bridge, other than a sketch in KGM archives.[21] It is a three-hinge arch with a 47 m span and 49.2 m total length. The total length of the bridge was increased to 49.20 m with 1.10 m cantilever extensions over the springing piers. The rise of the arch is 6 m and the width is 4.75 m between parapets (Figure 5.9).

The parapets of the bridge are made of concrete posts with guardrails in between, the posts are aligned with the spandrel walls. The approach parapets at the entrances are made with solid walls with a downward curved capping.

Irva bridge foundation system uses timber piles. Since the soil profile is soft clay, 75 wooden piles were used for each abutment (Figure 5.10).

The Contractor of the bridge is named as: "Türk Anonim Şirketi (Societe Anonyme Turque d'Etudes et d'Entreprises Urbaines Constantinople)" in KGM records. The original drawing[22] of the bridge was prepared using German and Arabic letters as the bridge was designed before the change of the Turkish alphabet.

From the drawings of the bridge, the corner of the abutment concrete walls was decorated with detail called quoin. This detail can also be seen in the Kirazlık bridge.

The construction of the bridge started in October 1925, and in June of the following year, the bridge was opened to service.

Figure 5.9 Irva bridge from Nafia album (İBB Atatürk Library).

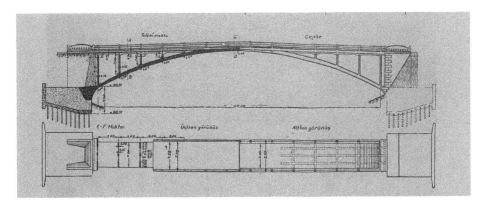

Figure 5.10 Irva bridge elevation and plan from Nafia Report (1933).

ACH-02: Kirazlık bridge

Irva bridge and Kirazlık Bridge are almost identical except for their parapets and foundation system.

This bridge on the Bartın-Safranbolu road crosses over Gökırmak river, which flows into Kızılırmak river. The concrete parapet chosen on this particular bridge is described by Nafia as: '...providing a more beautiful appearance...'.[23]

Bridge balustrade posts are aligned with the spandrel piers, and concrete balusters are placed between the thicker posts. The balustrade continues off the bridge for another 3 m.

The details at the corners of the abutment walls (quoin) are characteristic. There was also a date inscribed on the wing wall of the eastern abutment (Figure 5.11).

Kirazlık bridge is a three-hinged bridge, i.e. both the springing and the crown have been designed with a hinge, and these joints are free to rotate and do not carry the moment. The span between springing hinges is 47 m and the rise of the arch is 6 m. The total length of the bridge was increased to 49.20 m with cantilever extensions. The width of the bridge is 4.55 m between the kerbs and the total width of the arch is 5.05 m (Figure 5.12).

One of the abutments rests on the solid rock and its foundation is in the form of a spread footing. The other abutment rests on soft clay; its foundation was excavated with a concrete caisson, and 78 concrete piles were driven.

The tender for its construction was made in December 1926 and the bridge opened to service in September 1928.

The bridge will be submerged under the dam lake, which is named Kirazlık Köprü Dam. The dam is being built for energy generation and water consumption needs. It will also serve as a control measure for regulating the frequent floods in this region.

The construction of the dam was still in progress during the Author's visit in 2015. This bridge was found intact with no signs of deterioration in concrete. Only a continuous separation/crack in the transverse direction resulting from opening the hinge at the crown was observed during the site visit (Figure 5.13).

Figure 5.11 Kirazlık bridge photograph; the date inscribed on abutment wing wall (author's collection).

Figure 5.12 Kirazlık bridge from Nafia album (İBB Atatürk Library).

ACH-03: Akçay bridge

The bridge is on the Nazilli-Bozdoğan road in the province of Aydın. It is crossing over Akçay, one of the main tributaries of the Menderes River. The wooden bridge located here collapsed in 1923, resulting in the need for a permanent structure (Figure 5.14).

Figure 5.13 Kirazlık bridge (author).

Figure 5.14 Akçay bridge (KGM album 1988).

It has five spans of 35 m each. The spans are continuous deck arches with three ribs for each arch. The ribs are interconnected with struts, which are well detailed, adding positively to the appeal of the bridge. The depth of the section used for the arches is 60 cm at the middle spans and 40 cm at the side spans. The road width is 4 m with a 50 cm wide sidewalk on either side.

Akçay's riverbed can reach 400–500 m width in the vicinity; however, it decreases to 187 m at the bridge site. The speed of the current is 1.5 m/sec in normal conditions, increasing to 4 m/sec in extreme cases. Akçay normally has a low water level; however,

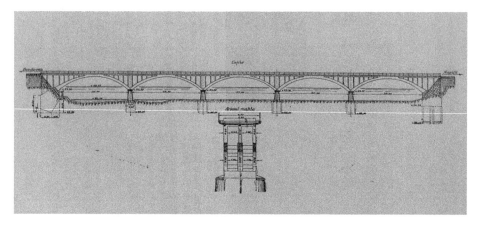

Figure 5.15 Akçay bridge elevation from Nafia Report (1933).

during the time of the flood, it has been determined that the water level can rise to 5 m or even 7 m (Figure 5.15).

For the planning of the bridge, it was considered that during the flood, trees and other debris are carried with the flow. Also, ground investigations revealed that the foundations needed to be lowered to a depth of at least 6 m. Therefore, to reduce the number of piers, the bridge spans were required to be as long as possible and the continuous arch system was chosen for this purpose.

The foundations were built with reinforced concrete caisson. The bridge was tendered on 23 June 1923 and construction finished on 14 April 1928.

A design drawing of the bridge is given in the thesis of Örmecioğlu,[24] who found it in the archives of KGM. This drawing is prepared using Arabic letters, as the bridge was designed before the change of the Turkish alphabet. This drawing shows slight differences with the constructed bridge, which is detailed as per the drawing given in the Nafia Report (1933). For example, the KGM design drawing shows the arrangement of spandrel columns to start as an extension of the pier, whereas the constructed bridge has the first spandrel column to start beyond the pier. The detail of the strut is again different for the mentioned two drawings (Figure 5.16).

Akçay bridge is a unique representative of the use of concrete deck arch bridge built with three ribs for the period.

ACH-04: Aslan bridge

Aslan bridge is located in the Zonguldak province, on the Ereğli-Devrek road. This exceptionally slender bridge has two spans, each of them 24 m. Each span has two independent ribs. The abutments and pier of the bridge are built on rock (Figure 5.17).

The bridge was tendered on 24 June 1927 and completed on 24 October 1928.

The original parapets on one side still appear in a recent photograph. The parapets are unique with architectural star-shaped detail in the middle. The bridge pavement has been filled excessively and almost reached half the height of the parapets.

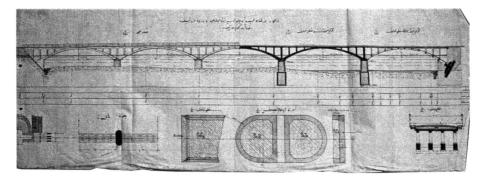

Figure 5.16 Akçay bridge drawing (KGM achieve).

Figure 5.17 Aslan bridge (author's collection).

ACH-05: Big Agonya bridge

Big Agonya bridge is constructed on the Balıkesir-Çanakkale road, crossing over Gönen Stream. As the ground conditions were good, a continuous arch type was chosen for the structure (Figure 5.18).

According to Nafia (1933): "The arches are concrete, the parapet is made with reinforced concrete, the pier and facade walls are stone". This means that, presumably, the arch itself did not have any reinforcement. However, the parapets of the bridge are heavily reinforced as can be seen from the current photograph of the bridge.

The bridge is of a solid spandrel type and has three spans each of 26 m. The width between parapets is 6.9 m. The total length of the structure is 105 m (Figure 5.19).

Big Agonya Bridge, together with Small Agonya bridge, are the only examples of this kind of concrete bridge in Turkey.

Figure 5.18 **Big Agonya Bridge (Ertuğrul Ortaç/www.ortac.net).**

Figure 5.19 **Big Agonya (KGM album 1988).**

Taking into account the possibility of flooding during construction, the scaffolding was supported with a temporary concrete pier constructed beneath the crown.

The arch section has a depth of 70 cm at the crown and 100 cm at the springing.

As mentioned in the Nafia Report (1933), the structural calculations for the continuous arch span were undertaken using a method known as "elastic theory". In this method, the arch structure is analysed by using numerical tables. This method was developed by German Engineer, Albert Strassner, whose book was translated into Turkish by Ali Fuat Berkman (Figure 5.20).[25]

Agonya bridges and Müstecap Bridge (BRD-18) were tendered to Hayri Bey with a contract dated 30 July 1930 and all bridges were completed on 25 November 1931.

ACH-06: Small Agonya bridge

This bridge was built similarly to the design of the Big Agonya Bridge with smaller spans of 8 m crossing over the Small Agonya stream on the Balya-Çanakkale road.

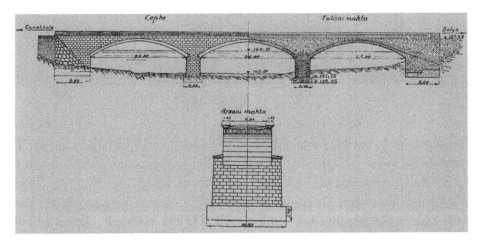

Figure 5.20 **Big Agonya from Nafia Report (1933).**

The three-span structure has the same width as the Big Agonya bridge of 6 m. Further information about its design details is not provided in the related documents. Its total length is approximately 50 m (Figure 5.21).

The bridge is a solid-spandrel type, in which the arch and spandrel facing walls were constructed and then the enclosed volume filled up to the deck level.

The foundations were excavated using wooden sheet piles. The ground is hard clay to an average depth of 3 m.

The bridge, which was tendered on 29 July 1930, was completed on 25 November 1931.

ACH-07: Bakırçay bridge

The bridge was on the İzmir- Bergama road and the location was specifically chosen where the riverbed is narrowed and limestone rock forms a firm embankment for the bridge (Figure 5.22).

This delicate bridge is a deck arch type with two segmental arch spans. The arches are three-hinged with spans of 26.6 m. The arch is a barrel type and supports the deck with spandrel columns.

Although the initial plan was to pile under the foundations and excavate the pier with wooden sheet piles, during construction it was decided to lower the foundations further so that the piles would be abandoned. The excavation was carried out with diaphragm walls at the abutments and a concrete caisson for the pier (Figure 5.23).

In order to reduce the weight and also gain in span length, a reinforced concrete deck slab was cantilevered at the springing. The deck rests on four longitudinal beams, which are supported on columns at every 3 m. The parapet is reinforced concrete.

The tender was awarded to SAFERHA with a contract of 9 November 1930 and completed on 24 November 1931.

No further information is found for the bridge but a possible location is given.

Balıkesir Vilâyetinde, Balya - Çanakkale yolunun 62+750 nci kilometresinde, Küçük Ağonya
deresi üzerinde 3 × 8 m. açıklığında, kârgir (KÜÇÜK AGONYA KÖPRÜSÜ) 1931 senesinde
yapılmıştır. Bedeli 25784,82 liradır.

Figure 5.21 Small Agonya bridge from Nafia album (İBB Atatürk Library).

İzmir Vilâyetinde, İzmir - Bergama yolunun 91+000 inci kilometresinde Bakırçayı üzerinde
2 × 26.5 m. açıklığında, betonarme (BAKIRÇAYI KÖPRÜSÜ) 1931 senesinde yapılmıştır.
Bedeli 45008.69 liradır.

Figure 5.22 Bakırçay bridge from Nafia album (İBB Atatürk Library).

ACH-08: Göksu (Beykoz) bridge

This bridge is in the Beykoz district of İstanbul, next to Anadoluhisarı (Anatolian Fortress) crossing over Göksu creek which flows to the Bosporus. It is so lost and crowded in by the buildings around it, that it is hard to realize that there is a bridge there. Although the nearby boats add to the scenery and can be considered a part of the image, the cables and pipes attached to the sides along the bridge disturb the view (Figure 5.24).

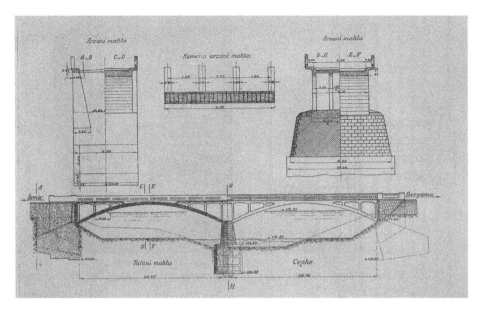

Figure 5.23 Bakırçay bridge from Nafia Report (1933).

Figure 5.24 Göksu bridge when completed with the remains of the older timber bridge alongside (FATEV archive).

The bridge replaced the former timber bridge. It can be seen from old photographs that the wooden bridge was still in use during the course of the construction of the new bridge. Also, the wooden bridge's decking and railings appeared to be in fair condition. Therefore, one can conclude that the reinforced concrete bridge was built in response to the demand of vehicle traffic (Figure 5.25).

Figure 5.25 Göksu bridge, South facade with Anadoluhisarı (Anatolian Fortress) at the background (author).

Figure 5.26 Göksu bridge, North facade original structure with a pipe attached under the parapet (author).

The bridge is the only known representative of the closed-spandrel arch bridge in Turkey. It was built as cantilevers extending from both banks and connected in the middle with a beam placed between the cantilever extensions. The two joint locations can be seen in the older photographs.

The northern downstream side facade of the bridge looks original, but since a large pipe was attached to it, the joints and the arch ring are not visible. Looking at the south facade, the original appearance of the bridge has disappeared. This view of the bridge suggests that the bridge may have been widened or otherwise modified because the arch and the second ring on the arch barrel (archivolt) have disappeared (Figure 5.26).

The local Municipality undertook the construction of this bridge; therefore, it could not be found in KGM records.

The completion of the bridge was announced in the newspapers: On 07 March 1932, the news announced the bridge was completed 2 months earlier and the transfer of the asset to the authority was accomplished. The news the next day on 08 March 1932 dated newspaper informs that the opening ceremony has been cancelled.[26]

The bridge span is measured as approximately 29 m long and 12 m wide from Google Maps.

ACH-09: Kömürhan (İsmetpaşa) bridge

Three Bridges born with the same name!

The Southeast region of Turkey is separated from the western part of the country by the Munzur Mountains and the Fırat (Euphrates) River. Transportation in the region was provided only by the İzoli (İzoğlu) ferry across the Fırat on the Malatya-Elazığ road.

The average width of the Fırat Riverbed in the vicinity is from 300 to 500 m and the water depth varies between 3 and 4 m during normal dry conditions. Moreover, the water level can rise up to 14 m during flooding, leading to very tough conditions for transportation services. Thirteen km downstream from the İzoli ferry crossing is the location called Kömürhan, where the river width reduces to 100 m. On the other hand, the lowest water depth is 12 m and during flooding the depth increases to 20 m carrying debris with the strong current, thus making the construction of bridge pier foundations within the river unfeasible (Figure 5.27).

Although the construction of a permanent bridge was planned for a long time, it was not realized for many reasons. However, a 543 m long timber bridge was constructed

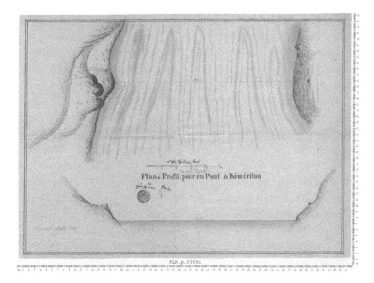

Figure 5.27 **Preliminary studies of Kömürhan bridge (BOA PLK-p-3591).**

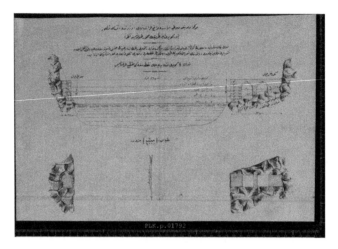

Figure 5.28 Preliminary studies of Kömürhan bridge (BOA PLK-p-01792).

by soldiers for military access during the independence struggle. This bridge required repair nearly every year, and the transportation was frequently interrupted.

In 1926, a design for a steel bridge was prepared for the Kömürhan location, as it would be hard and very expensive to span the crossing with a concrete bridge. This steel bridge was designed with screw piles. However, the design failed to be further pursued due to the lack of budget (Figure 5.28).

Preliminary designs were also prepared by SAFERHA company in 1928 for a single-span bridge at three possible locations. The sketches for the bridge were shown together with detailed construction staging with temporary towers and cable systems.[27]

The wooden bridge was swept away during the flood of 22 April 1929. Therefore, the need for a permanent bridge became unavoidable. The decision to construct the bridge was signed on 12 March 1930.[28]

The Fırat's water level was predicted to rise 7–8 m at İzoli. However, a flood in April 1929 showed that this level can increase by up to 14 m. After experiencing the rise higher than predicted levels, it was clear that the bridge would be more expensive than initially assumed if it was to be constructed at İzoli. In addition to the bridge, the road on Elazığ side would have to be elevated in the approaches above its existing level along with the 13 km long distance where the flood basin expands. Considering these factors, it was better to construct the bridge at the Kömürhan location.

Both embankments at Kömürhan valley are rock and steep; therefore, an arch bridge was the first choice. Field measurements revealed that the minimum arch span would be about 105 m. This bridge could be made of concrete or steel. Although for concrete, it is not a serious concern to be submerged in water, for steel, this should be avoided for durability. Therefore, the steel option required a longer span to keep the whole arch free from water interaction. In addition, the regular maintenance and painting requirements would be very hard for such a large and not easily accessible

Figure 5.29 Kömürhan bridge when completed 06 October 1932 dated archive (BCA 155-90-8).

steel bridge; therefore, the decision was on a concrete arch even though it would be higher in initial price (Figure 5.29).

Scaffolding: The flow speed, flow depth, expected flood levels and the depth of the gorge made it practically impossible to construct a pier, resulting in a real challenge for constructing scaffolding. The scaffolding would basically have to achieve the same span as the bridge.

Scaffolding at that time used to be made either from metal or wood. In either case, it could be constructed on-site or elsewhere, and be transported to the bridge site. Other options could be a progressive construction of cantilevers extending from each bank of the river or supporting cables stretched between the shores of the valley.

An alternative, in case of metal scaffolding, after the construction it can be left in place to serve as a permanent reinforcement integrated into the concrete structure. This type of scaffolding is known as "Melan", after the Engineer Josef Melan who was the inventor of this system. For this alternative, the reinforcement made of steel is used as scaffolding and at the same time it is supporting its own; after concreting, it stays in the concrete, acting as reinforcement. Even though this system was easy to assemble, it was expensive at that time in Turkey.

Tender: Usually, Nafia developed the designs for the proposed bridges and left the construction stage designs to the Contractors. For this tender, the candidates were able to propose alternative designs abiding by the conditions outlined in the tender specifications.

One of these specified conditions was on the type of the arch; if the deck adopted a rigid type, i.e. deck-stiffened arch, then the failure of one spandrel column would cause the whole arch to collapse. Therefore, the condition for the main arch to be designed to carry the moment and axial forces by the support of a full-width barrel or two ribs was specified in the tender documents.

Two tentative locations 50 m apart, were offered to candidates for tender. They were free to select either of these locations in their proposal.

Two participants were eligible to meet the conditions of the tender. These were Sweden's Nydqvist-Holm[29] and Denmark's Christiani-Nielsen.[30] The Nydqvist-Holm (NOHAB)[31] presented the lower-priced proposal. Thus, the construction of this bridge was awarded to them, and the contract was signed on 26 July 1930.

NOHAB proposed the use of a wooden scaffolding system, similar to the one that they used for La Caille bridge in France. Whereas, the Christiani-Nielsen company had proposed a Melan steel scaffolding system.

La Caille Bridge was a 137.5 m span crossing over a deep ravine in the Alps and was constructed with wooden arch centring suspended from towers. The bridge was designed by Albert Caquot (1881–1976), a prominent engineer, and construction was carried out by Compagnie Lyonnaise d'Entreprises in 1928. This method had been proposed earlier by SAFERHA in their 1928 design.

Structure: The main arch span had a 109.60 m structural length with a height of 23.9 m. The total length of the bridge was 164 m including the four approach spans, two at either end, of 12 m each (Figure 5.30).

The main arch segment was a concrete box section of two cells with the dimensions of 4.80 × 1.40 m at the crown and 6 × 2.26 m at springing level. The bridge deck constituted a 4.8 m road with 80 cm wide sidewalks on both sides. The overall width was 6.4 m.

The finished surface level had a slope of 3% longitudinally symmetrical about its centre; in the transverse direction, the deck had a slope of 2.5%. The deck slab is resting on the transverse beams which are spaced at 5 m intervals. Transverse beams are supported by spandrel columns.

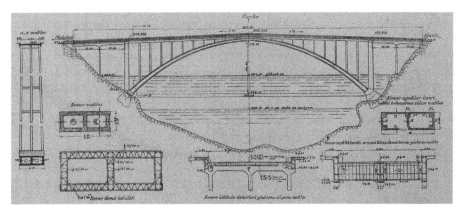

Figure 5.30 Kömürhan bridge details from Nafia Report (1933).

The design was made in accordance with the 1925 dated German Concrete Standard. Moreover, the advances that had since occurred in cement production at that time were taken into account and the concrete allowable stress was, therefore, allowed to reach 80 kg/cm² for arch and 50 kg/cm² for the slab.[32]

In order to keep the scaffolding structure as light as possible, the concreting of the structure was done in stages. In the first stage, the scaffolding only carried its load and the bottom flange of the arch. In the second stage, the webs were poured and the weight was carried by the scaffolding and the bottom flange of the cross-section. Lastly, the upper flange was concreted and carried by the rest of the cross-section concreted before.

Sand and gravel required for the manufacture of concrete were taken from the Kömürhan creek which discharges to Fırat. Arch concrete is cast with "super cement" from the Dyckerhoff Doppel brand, and for the other parts of the structure, cement from the local Yunus cement factory in Kartal was used. Kömürhan was the first bridge in Turkey to be constructed with the high-strength concrete also called "super cement".

Concrete Tests: Since the concrete capacity was pushed beyond its limit, the construction specifications required that a special testing regime had to be carried out; for every 100 m³ of concrete, six testing specimens were to be manufactured; three tested after 7 days and the remaining three of them were tested after 28 days. The test results, according to German concrete specifications, are divided by a factor of 1.7 to determine the allowable capacity for beams. NOHAB's documentation[33] provides some explanation of concrete testing procedures.

Towers: In order to construct the scaffolding, the approach spans and the pier at the arch springing were constructed first. Wooden towers were then erected on top of the piers at the springing. Then four cables were stretched over these towers, each with the capacity to carry up to 14,700 kg. The cables were anchored to strong bolts anchored into the rock with concrete at both ends of the bridge.

Scaffolding was made of five wooden truss arches placed at intervals of 1.25 m. Each arch was made of 33 identical elements joined to form the curve. The depth of the scaffolding arch was 3.5 m constant along the entire curve (Figure 5.31).

An interesting account of the scaffolding erection can be found in the Bauingenieur Magazine.[34] The scaffolding was built with lumber from the Taurus region, which was so dried out that a strict supervisory service was put in place for preventing fire, and a smoking ban had to be issued to prevent an accidental conflagration.

As the scaffolding was completed, the hinges at the crown and springing were also filled with dry mortar so that the scaffolding would work as a fixed arch for the upcoming construction loads. The crown was also secured with cables tied to the foundations on the banks. A test loading of 10 tons of cement was placed at the crown to confirm the required strength.

Opening Ceremony: The completion of the bridge was reported to the Prime Minister in June 1932. Atatürk named the bridge "İsmetpaşa Bridge".[35] The opening and naming ceremony was held on 5 October 1932, with the participation of Prime Minister İsmet İnönü, who was born locally in Malatya. Both ends of the bridge had a ceremonial arch which were established by the provinces of Elazığ and Malatya. The arch on the Elazığ side was "white and red" ornaments, matching the colours of the national flag. The ribbon was stretched at the entrance and

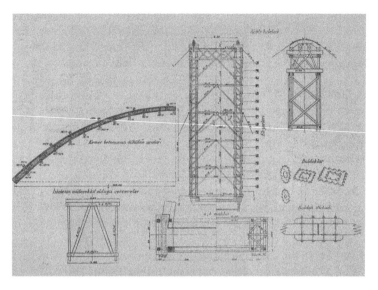

Figure 5.31 Kömürhan bridge scaffolding tower details from Nafia Report (1933).

Figure 5.32 Kömürhan bridge opening ceremony (STFA archive).

a sign plate with text in large font written "İsmetpaşa Bridge" was placed on the portal (Figure 5.32).

Fırat river was an obstacle for the people in the region for many years. The crossing was very challenging with floods and raft transportation was very prone to accidents. Many lives were lost to the river and a response to this was made in a form of "writing laments" to express the anger and sadness of the people. Hence, the bridge was a landmark and made a significant contribution to the lives of locals.

One of the folk songs from the region tell the story of many sorrows:

> Kömürhan Bridge faces Harput / Cruel Fırat, to be blind, destroys families/ Don't make me talk / I have a deep sore [36]

Kömürhan was a technological crown for the new Republic. Turkey gained an international reputation in the field of construction and engineering of complex structures for the first time with this bridge. When Kömürhan was built, the reinforced concrete arch bridge was the seventh longest span ever built.[37] Even though the design and construction were headed by foreigners, the local authority was fully involved with the development of the project and had control over the decisions regarding the project.

Kömürhan can be considered a turning point for the bridges built in the country later on. It has been in the articles of newspapers with headings such as "The large concrete bridge which comes first in Asia"[38] and "The bridge policy gives good outcomes in the country: How the Bridges that could not be built for centuries brought to life in 3 years?",[39] the latter article announced the name of the new bridge as İsmetpaşa Bridge (Figure 5.33).

Kömürhan Today: Kömürhan bridge is unfortunately no longer visible as it is fully submerged in the lake of Karakaya Dam.

The 1932 bridge was superseded by a balanced cantilever bridge with a 135 m span and 60 m height, which was the first of its type in Turkey when it was completed by STFA in 1986. Later, in 2012, a third bridge with the name Kömürhan was completed, it was a cable-stayed bridge with 380 m span crossing over Fırat and the first two bridges (Figure 5.34).

Figure 5.33 Title of the articles reported for Kömürhan bridge opening, left: 1 June 1932 and right: 15 August 1932.

Figure 5.34 The two Kömürhan bridges, 1986 Cantilever Bridge at the background (KGM album 1988).

ACH-10: Nişankaya bridge

This bridge is located over the Sarıcaeli Creek on the Çanakkale-Balıkesir road. It has two spans of 18 m each (Figure 5.35).

A continuous arch was chosen as the structural type due to appropriate ground conditions and sufficient headroom. All the foundations are resting on rock. The river stream can potentially carry large rocks of up to a quarter cubic meter in volume during a flood, therefore the pier was protected by stone riprap.

The section of the arch member is 30 cm deep at the crown and 48 cm at springing, the arch width is 5.40 m.

Nafia reports that the continuous arch spans were calculated according to the "Strassner method", in which numerical tables are used to analyse arch structures (Figure 5.36).

Figure 5.35 Nişankaya bridge from Nafia album (İBB Atatürk Library).

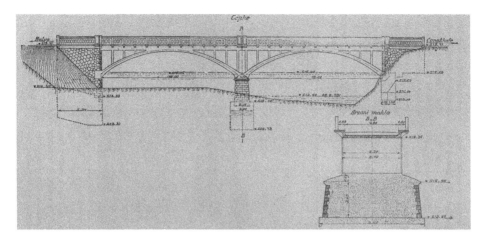

Figure 5.36 Nişankaya bridge from Nafia Report (1933).

The deck rests on the reinforced concrete walls spaced 1.5 m apart. The road width is 4.80 m with sidewalks of 60 cm wide on both sides. Stone paving was used on decking and the concrete balustrades with posts at pier, abutments and entrances.

The construction commenced on 20 March 1932 and was completed on 30 October 1932.

This charming bridge still preserves its original features like abutment wall finishes, and especially the original balustrades remain intact all along the bridge. It is abandoned by the current road network.

ACH-11: Fevzipaşa (İspiroğlu) bridge

This bridge is in the Sinop province, on the Boyabat-Kastamonu road, crossing over Gökırmak, one of the branches of the Kızılırmak river. At this location, an old bridge with the name İspiroğlu had collapsed; therefore, a new replacement bridge was needed (Figure 5.37).

There are two identical spans of 26.76 m each, similar to the Bakırçay Bridge in İzmir. The abutment on the north side was excavated and constructed during the dry season. The abutment on the south side was excavated using diaphragm walls. The pier excavation was made with wooden sheet piles. The foundations are made with masonry using cement mortar. The reinforced concrete elements were made with 350 doses, i.e. 300 kg cement used in every m^3 of concrete (Figure 5.38).

The width of the arch is 3.3 m and the depth of the section was variable with 35 cm at the crown, 52 cm at the haunches and 40 cm at the springing. The road width is 3 m with 60 cm sidewalks on both sides. The deck rests on transverse beams which are supported by columns spaced at 3 m. The concrete balustrades are formed by rectangular panels.

The construction contract is dated 24 March 1932 and was completed on 9 January 1933. The acceptance of the bridge was made on 4 June 1934. The Contractor of the bridge was a company called Ankara İnşaat which had three founders Mehmet Galip, Fesçizade İbrahim Galip and Erzurumlu Nafız (Kotan) (Figure 5.39).

Figure 5.37 **Fevzipaşa (İspiroğlu) bridge (SALT research, photograph and postcard archive).**

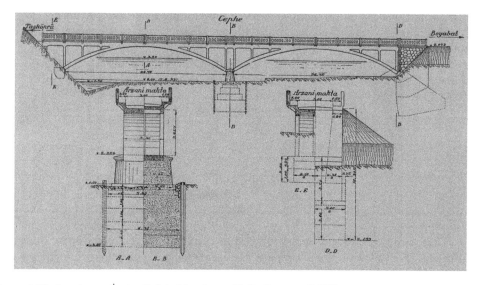

Figure 5.38 **Fevzipaşa (İspiroğlu) bridge from Nafia Report (1933).**

Sinop province is connected to the rest of the country with this bridge; otherwise, it would remain isolated. Upon the desire of the local people, the bridge was named after the General of the Turkish Armed Forces of the time, Fevzi Pasha.[40]

This bridge is known as "İspiroğlu" in the memoirs of Akkaya[41] who was working at Nafia during this time[42]:

Chief (Kemal Hayırlıoğlu)[43] took me to Anatolia on my first trip. I guess this was a training trip... At dawn, we set off from Ankara with a rental taxi...

Figure 5.39 Fevzipaşa (İspiroğlu) bridge during the handover of the bridge (FATEV Archive).

.... The next day we crossed Taşköprü (Kastamonu) and entered the Gökırmak Valley. All these places we passed by, have not been under enemy boots, also have not been touched by friendly hands.

.... within 20 km to Boyabat, the chief left me at the construction site of İspiroğlu Bridge. I liked the instructions he gave. "What you see is true. Make them correct, let them finish the arch concrete works, then leave!"

The construction site chief, a random German who was a bridge lover (referring to the game called bridge) and Ali Çavuş who was indeed from the İspiroğulları (a family in the region), were on the construction site. I stayed during a month at the construction site. We played bridge every night and ate hunting meat every day...

Later, Akkaya was also sent for the handover of the bridge:

...At the beginning of winter, the İspiroğlu Bridge construction was finished, and I was sent to the acceptance of the bridge accompanied by Seyfi (Tonga), however Seyfi Bey was appointed as a lead... When I became aware of his title, I concluded that I was somehow punished despite Seyfi Bey being older; because I was the one experienced with this bridge.

When we met with my old friend Gökırmak [River], we found the Sinop Nafia Manager ready at the bridge location. He was an old man dressed in black, with a fedora, and still strong enough to shake hands. Seyfi Bey was very reluctant. Although we, as the side with the fedora hat,[44] objected, Seyfi Bey was stubborn in load testing. İspiroğlu Ali Çavuş's tumbrel groaned and pulled tons of stones and stacked them on the arch. The bridge was therefore closed for three days. During all these three days, all travellers had no option but to cross through the river by striking their donkeys.

ACH-12: Körkün bridge

Körkün bridge was built on the Adana–Karaisalı road over Eğlence River, which is a branch of Seyhan River. The bridge was tendered to SAFERHA on 26 October 1932 and completed on 7 February 1934.[45]

During the preliminary investigations, a difference of 45° was observed between the normal and flood flow directions of the stream at the bridge site. This meant that the pier had to be exposed to oblique – non-axial – currents during flooding. Therefore, a single 72 m span bridge avoiding piers was proposed by the Contractors and accepted by Nafia (Figure 5.40).

The bridge is a half-through type with arches interconnected by lateral beams forming K-shape on the plan. The 70 cm wide arch section had 120 cm depth at the crown and 196 cm at the springing. The width between the arches was 4 m.

Bridge is designed by Halit Köprücü,[46] who described the structure in an article as[47]:

> … one of the most mature works of the early Republican era, and an important symbol of both the technical knowledge and the workforce of Turkish engineers. It has a special value in terms of demonstrating that they have never lagged behind their foreign colleagues in this regard.

The construction of the main structure was finished by working day and night. After construction was finalized, it was left for one month for the hardening of the concrete to remove the temporary formwork.

The application of formwork removal by the Freyssinet[48] method was first experienced in this bridge, in regards to road bridges in Turkey. This method had earlier been applied to railway bridges in Turkey.

In the Freyssinet method, the two halves of the rib are left uncompleted at the crown during concreting. The halves are then further separated against each other

Figure 5.40 A card sent to Akkaya reads 'the work praises the master' (FATEV archive).

by jacks. This jacking force applied to the arch lengthens the axis of the ribs, coun-teracting the rib shortening and shrinkage stresses which occur in concrete. The force applied results in separating the two halves at the crown and lifting the entire rib from the formwork. High-strength concrete is then placed in the opening to close the crown and interconnect the rib halves to form the arch.

Hydraulic jacks with a capacity of 100 tons were placed at the crown and each half arch was raised by the pressure applied on them, allowing the formwork to be easily removed.

The self-weight of the bridge was transferred from the scaffolding to the arches by lifting the arch at the crown opening the arch rib for only 10 mm and later closing this gap by grouting. This reduced the adverse stresses caused by shrinkage and dis-placements from the self-weight of the arch. This process was carried out by exactly following the methodology and calculations which were made well in advance, without noting any setback or unexpected result.

In addition to the technical benefit of following this process, the bridge structural dimensions and the reinforcement amount were significantly decreased, thus, reducing the construction cost of the bridge (Figure 5.41).

Akkaya was working in Nafia as a graduate engineer[49]:

The 72 m Körkün Arch Bridge in Adana and the 50 m Pasur Bridge projects of Sezai (Türkeş)[50] were assigned to me. These were great bridges at that time...

He (Chief Kemal Hayırlıoğlu) sent me to the project in Adana to "have a look" at the foundations of the Körkün Bridge, which was entrusted to me. I went by train. It was the first time I was passing the imposing Taurus (Mountains). You can imagine my puzzled self. During the 15 days I stayed in Adana, I met Halit

Figure 5.41 **Körkün bridge is ready for opening (FATEV archive).**

Köprücü, who would later become my boss for many years. It was then when I understood the first concepts about the "Formations of Çukurova (Region)" that I will come across many times in the following years. Finally, I had the foundation concrete poured…

(Error!) On my return to Ankara, the chief told me off thoroughly. I did not realize that "concreting the foundation" was one thing, and "looking to the foundation" was another. It turns out that, the concreting for the foundation was not a decision that was supposed to be made without spending an average investigation period of 1.5 months…

The reason for the faulty record of an inexperienced professional like me was that I used a mix of my instinct and the experience of Halit Köprücü…

After that, Kemal Bey did not send me for inspections anywhere for a long time.[51]

Akkaya made an early decision about the concreting the foundation of the Bridge and was left practically idle as a punishment for a while by his direct manager.

The bridge completion was reported in a newspaper dated 15 February 1934[52] stating that "Another Reinforced Concrete Bridge which is exceptional in Europe was constructed", providing details for the crown opening during the construction (Figure 5.42).

Nowadays, the bridge is submerged in the Seyhan dam lake reservoir.

ACH-13: Pasur bridge

This bridge was built by SAFERHA. There is an article written in Arkitekt magazine by Halit Köprücü, and the name of the Chief Bridge Engineer Kemal Hayırlıoğlu was also added to the title.[53]

Figure 5.42 Körkün bridge in the newspaper described as an "Exceptional type in Europe".

Figure 5.43 Sadık Diri and Sezai Türkeş during the construction of Pasur bridge (FATEV archive).

Pasur bridge was planned during the independence war. In 1928, the Province tendered this bridge to the contractor; however, the contractor could not find an engineer to manage the site works and the contract was cancelled (Figure 5.43).

The Bridge crosses over Pasur stream, a branch of the Tigris, 20 km from Siirt on the Siirt-Diyarbakır road. The bridge site is located where the stream suddenly enters a narrow gorge from a wide valley, causing the water flow rate to be high.

Türkeş[54] provides some information about the conditions experienced: "since the road was not suitable for the tumbrel travel, the cement and other material was carried and transported by mules and camels to the construction site. Since camels were unable to carry a concrete mixer, the preparation of concrete had to be done manually on site".

The bridge span is 50 m with a rise/span ratio of 1/6.5 and with a width of 3.7 m. The foundations rest on solid rock. Stone Parquet pavement was laid over the deck.

The main challenges encountered during the construction of this bridge were listed by Halit Köprücü as the stream bed was being mainly full of gravel, resulting in the foundation being exposed to deterioration by scour and the water flow rate being very high as the riverbed narrows at the bridge location and the high probability of flood, especially in autumn. In addition, timber was scarce in the vicinity and the only type available was ordinary poplar. Therefore, timber required for the construction had to be transported from Adana.

In order to reduce the amount of timber needed, the temporary scaffolding was designed as a truss in the form of an arch.

The design assumptions and design calculations were verified by testing made on the scaffolding during its construction. Results were consistent as intended.

During the placement of reinforcement on the formwork, a flood occurred and the water rose to 50 cm below the highest water level ever recorded, washing away the temporary timber service platform. Although there was not much weight supported

Figure 5.44 **Pasur bridge after its completion (FATEV archive).**

by the formwork, which would indeed increase the stability of the arch, no detrimental damage happened to it (Figure 5.44).

Details of Arch Scaffold: The truss arch consists of solid planks of 4.5 × 20 cm. The upper and lower chords are composed of four rows of planks that are nailed against each other. These planks were given a curved shape to fit the intrados – lower soffit – shape of the bridge arch. There were four truss ribs spaced at 1.16 m. After the scaffold was prepared on the ground, it was lifted and installed with the assistance of an auxiliary pier placed under the crown.

Concreting: The concrete was poured in two stages and each layer was divided into three sections. First, the middle part of the lower layer, then the side sections poured in one day, and after a one day break, the upper layer concrete was poured in the same order. The whole section was finished by pouring concrete simultaneously so that the load on the arch balanced symmetrically.

In order to minimize the stresses resulting from concrete shrinkage and temperature changes, the concreting was planned to be undertaken in cold weather and scheduled for autumn.[55] The construction was started towards the end of August 1933 and the arch was ready for concreting with the reinforcements placed in the formwork by the following December. High-strength concrete was used in all reinforced concrete elements.

Because of the cold winter, it was not possible to work all day, and the concrete operation was limited the better weather, between 9:00 am and 2:00 pm. Although the weather temperature drops below zero at night during this time of the year, the concreted sections were covered with sacks and tent cloths, and together with the heat of hydration of the newly poured concrete, the temperature in the concrete

MEMLEKET HABERLERİ

Dördüncü Büyük Köprü De
Muvaffakiyetle ikmal Edildi

Figure 5.45 15 April 1934 dated Son Posta newspaper announces the Pasur bridge as
"fourth biggest bridge" upon its completion.

remained above zero. Measurements were checked regularly with thermometers placed under covers.

After pouring the lower layer concrete, a 2 cm sag was observed in the permanent structure and a further 2 cm was perceived with the completion of the second layer, totalling a sag of 4 cm at the crown. These measured values matched the predicted deflections in the engineering calculations.

The bridge was tendered on 27 March 1933 and finished on 5 April 1934.

Akkaya was working in the SAFERHA office while Türkeş was on-site with Sadık Diri. Türkeş also wrote to Akkaya[56]: "On the opening ceremony of the bridge, the scissor to cut the ribbon, fell inside the lining of the Governor's jacket, whilst his hand remained hanging in the air".

When completed, it was the fourth longest span of Turkey, and the news reported with the heading "The fourth longest bridge was constructed with success".[57] (Figure 5.45)

ACH-14: Göksu bridge

The bridge was built on the Göksu river, a tributary of the Fırat, to connect the Adıyaman-Besni-Gölbaşı road, with the Fevzipaşa-Diyarbakır railway line. The width between the parapets is 5.2 m, the arch spans are 34.6, 35 and 34.6 m, resulting in a bridge of 112.4 m length.

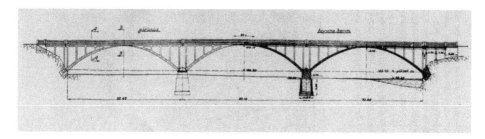

Figure 5.46 Göksu bridge drawings from public works magazine (İBB Atatürk Library).

The abutments rest on rock and the boreholes during ground investigations showed that a suitable ground could be found at a shallow depth. Therefore, an arch bridge with three spans made of reinforced concrete was the preferred solution. The rock was found at about 3–4 m at both piers and all the foundations were supported on rock. The pier excavations were made with concrete caissons (Figure 5.46).

As timber was not available in the vicinity, both timber and other construction materials had to be transported from Adana to Gölbaşı station, and then travel another 100 km on often very bad local roads to the construction site.

The scaffolding of the arches is in the form of a wooden truss arch to reduce the amount of timber needed due to the difficulty in transporting the construction material.

The bridge was tendered on 4 February 1933 and completed on 24 March 1935.

ACH-15: Güreyman (Kız) bridge

This elegant bridge is an 18 m span deck arch bridge with two ribs. It is very slender and it was found still in good condition, crossing a small creek at the 86th km along the Ankara-Beypazarı road.

"Kız" meaning "Girl" was the way the people named the bridge after the female engineer, Sabiha Rıfat (Ecebilge) Güreyman who worked on the construction of the bridge.

Güreyman entered the Engineering School in 1927 and graduated in 1933. She was one of the two female engineers who enrolled in the school for the first time in Turkey by special orders made by Atatürk (Figure 5.47).

She was appointed to Ankara Nafia (Public Works) after graduation. Güreyman took part in the construction of the bridge and worked as a controller on-site during 1934–1935.

Her interview notes had been published in a newspaper[58] and transcribed below.

Winter was coming. The roads were bad. The job had to be finished as soon as possible. If the snow hit, the work could be put on hold for two to three months. The workers were restless as well. They wanted an increase in their salary, but the contractor was not able to afford it.

Figure 5.47 Kız bridge view (Photograph: Mehmet Emin Yılmaz).

That night, Sabiha Rıfat had eaten and retreated to her tent. The contractor came in suddenly:

> They (workers) are leaving, lady engineer!
> I jumped out as I grabbed the kindling. I ran, ran. My thin voice echoed in the dark of the night. I remember with all my might I shouted "Stop, stop". I said, "Where are you going?" They replied that they would be able to earn more money somewhere else, so they went to the construction site of the mosque in the same village. Winter had also arrived, so they were afraid if the snow was to set in intensely. I said "Shame on you! Be ashamed of my femininity!" This bridge needs to be built more than a mosque, let's go back and finish our job. They talked between themselves and I said "We started together, we finished together". Later, the bridge was finally completed. I was able to learn years later that the governor heard about such events and was proud about my leadership. They called the arch bridge the "Girl's Bridge"... I was so happy when I heard this...

ACH-16: Meşebükü (Bolaman) bridge

A bridge awarded with a price increase!

Usually, in a competitive environment, tenderers are forced to make discounts, sometimes up to 30%, to win the tender. In contrast, this bridge was contracted to SAFERHA with a 12% addition in the estimated cost on 14 September 1933. This was probably because no company showed interest in the project which was announced on 24 August 1933[59], and SAFERHA was already in the region engaged for the construction of other bridges.

This bridge is located in the populated coastal zone of the Black Sea region and connects the coastal settlements of Ordu, Fatsa, Ünye and regional towns in the area which were limited to sea transportation until 1935 (Figure 5.48).

Until then, ferries would stop at the jetties on certain days of the week; however, they would not stop at every jetty. Besides, the jetties were unable to be used during

Figure 5.48 Meşebükü bridge from south upstream (author).

stormy weather. Therefore, these coastal settlements had to be connected to a safe harbour, Samsun, with a reliable and continuous road along the coast.

In terms of its geography, the Black sea region typically has mountains rising suddenly from the coast, leaving very limited space for housing settlements, and many rivers flow directly from these mountains to the sea. Since the region also has a very high rainfall intensity, streams can carry a lot of water during flooding times and, therefore, bridging these streams was very challenging.

The construction of a bridge over the Bolaman, which is one of the large rivers in the region, was originally planned with a 300 m wide crossing. Because it would be very expensive to construct a bridge of this length, a more suitable location was found approximately 4.5 km south inland from the coast (Figure 5.49).

The Bolaman Bridge is a half-through arch bridge with a 65 m span. It had one active hinge at the crown during its construction and this hinge was removed later once the bridge was finalized.

For the removal of the hinge, the two halves of the arch were moved away from each other by the application of a horizontal thrust exerted by flat jacks. Four jacks were used to open the crown to 14.5 mm using 350 tons of force, then, this gap was filled with a high-strength cement mortar and once it hardened, the hinge was closed.

Although the riverbed at the proposed bridge location was 285 m wide, the low water level was constant and was only 70 m wide during normal annual flows. Only at 5- to 10-year intervals, the flood would cover the entire riverbed. Therefore, on the west side of the entrance, the road was lowered to the level of the stream with a 12% slope and a 65 m span was considered enough (Figure 5.50).

Due to the very high probability of flood during construction, the arch scaffolding was built as a wooden latticed arch which was supported by cables from wooden auxiliary towers (Figure 5.51).

Figure 5.49 **Meşebükü (Bolaman) bridge section (author).**

Figure 5.50 **Meşebükü bridge from south-east (author).**

Figure 5.51 Meşebükü (Bolaman) bridge during construction (FATEV archive).

At the end of 1935, the construction of the bridge was completed except for the stone pavement. Testing and acceptance procedures were carried out on 8 November 1935 and the bridge was completed at the end of March 1936.

A document[60] in archives reveals that an inspection of the bridge carried out on 05 April 1936 found a deviation of 0.0015 in the alignment of the bridge. The deviation was larger than the maximum allowance specified in the contract, which was 0.0006 of the span. Therefore, the contractor received a penalty of 1500 lira, which had to be deducted from the overall contract price of 42,000 liras. This meant that the authority recovered 3% back from the original expensive contract.

ACH-17: Afrin bridge

This bridge was built on the Kilis-İslahiye road over the Afrin Stream, connecting the southernmost border city, Kilis, to the İslahiye railway station. The single-span bridge has a 36 m span and a total length of 54 m. Bridge width is 4 m in total with 50-cm-wide sidewalks on both sides. The section of the arch depth is 40 cm in the crown and 70 cm at the springing. The rise/span ratio of the arch is 1/6 (Figure 5.52).

On 31 December 1930, a decree was signed to issue a foreign exchange agreement to purchase materials from abroad for the purpose of construction of this bridge.[61]

The construction of the bridge started with the contract dated 28 February 1936 and was completed on 16 May 1937.[62] The contractor of the bridge was named as Fenni İnşaat Şirketi- İzmir (Eng. Ali Nihad and Ömer Lütfi) in KGM records.

ACH-18: Fadlı bridge

The bridge is in Tokat Province and crosses over Kelkit River on the Niksar to Reşadiye road.

Figure 5.52 **Afrin bridge from postcard collection (İBB Atatürk Library).**

Figure 5.53 **Fadlı bridge (author).**

It has a 36 m single span. The arch foundation rests on solid rock on the Niksar side and compacted sand pebbles on the Reşadiye side (Figure 5.53).

The superstructure design and details are identical to the Afrin Bridge. Since it was not feasible to drive piles into the stream for a temporary scaffolding, the construction was undertaken with a wooden truss arch scaffold.

Construction was tendered on 18 December 1935 and was completed on 24 March 1937.

This bridge and what appears to be the original balustrades appear to be in good condition as confirmed in the 2020 site visit made by the Author.

ACH-19: Keban Madeni bridge

Keban Madeni bridge was needed to connect Elazığ District with the towns of Arapkir, Kemaliye and Divriği. It was crossing over the Fırat (Euphrates) River.

The location of the bridge is on the Elazığ - Keban road, 3 km from the centre of Keban. The site was chosen for its narrow strait of the Fırat, and 2.5 km of road was constructed along the riverbank to reach it.

Since both banks were steep, the single-span arch system with a span of 62 m and a height of 13.5 m resulting in a ratio of 1/4.5 was preferred. The total length of the bridge is 112 m.

Since the volume of traffic crossing the bridge was low, the width of the bridge is only 3.7 m, allowing only a single lane of traffic with sidewalks of 40 cm on both sides.

The bridge is currently submerged under the dam lake of the Fırat (Figure 5.54).

Scaffolding: In an innovative method, each half of the timber scaffold was assembled in a near-vertical position on the river bank, and then rotated down to its intended place. The construction of the bridge is explained in detail in a magazine.[63]

The height from the crown to the lowest water level was 24 m, and the flow rate was very fast because the riverbed at the bridge location narrowed down to 50 m instead of the regular width of 200–300 m of most of the Fırat River.

As it was not, therefore, possible to erect any auxiliary structure supported through the river, two-hinged trussed-arch supported on the rock on both riverbanks were adopted as scaffolding. The scaffolding arch span was 50 m. It was designed for the self-weight of the fresh concrete of the arch with an additional weight of 200 kg/m² taken into consideration.

The arch-truss members were made of 4 m long timber planks sized 12 × 24 cm, connected by bolts and washers. In this method, uniform tightening was ensured by grouting the joints. The truss-arch was divided into ten total parts, each part was constructed at an area of 100 m above the Elazığ side and transported to the

Figure 5.54 Keban Madeni bridge (İBB Atatürk Library).

riverbank via a cable line. Parts were then assembled on the vertical half-arches, standing on the springing of the scaffold.

The scaffold system consists of four truss-arches, which are interconnected by lateral beams.

After the half trusses were strengthened with transverse and lateral beams, the remaining parts of the two middle trusses were placed. Half trusses with a height of 31 m stand vertically on skewbacks. Then, the truss- arch was slowly lowered. As a result, the middle truss bays are linked together in the crown. The same was repeated for all four scaffolding truss-arches.

The arch concrete was poured in one layer from the springing to the second uprights and the remaining parts to the crown in two layers, all subsequent parts were divided into sub-panels and finished in 20 days. After the arch concrete was poured, 3 cm deflection was measured at the crown. Twenty days after the last layer was poured, the wedges were removed and the arch was released from the scaffold. With a later measurement, no further displacement was noticed at the crown.

Due to the very steep riverbank on the Elazığ side, a construction activity platform was assembled 100 m above the bridge on the Elazığ side. Timber, reinforcement, cement, sand and gravel were stored on this site until the road connection was completed. After the reinforcement and timber members were prepared, the construction material was passed to the opposite shore with the aid of a rope line installed in the air. However, after the scaffolding was completed, it was possible to establish a storage area on the opposite shore.

This region has all kinds of difficulties: the steepness of the site, the narrow access road and constant winds, as well as the unprecedented heavy winter of 1936 and 1937, all made the work very challenging, as also noted by Nafia:

> However, it is appreciated that the contractor sorted all kinds of difficulties, but with patience, diligence, and the simple tools available at the site made an important contribution for the achievement of such as significant outcome on the Fırat.[64]

The bridge was tendered on 2 September 1935 and completed on 3 September 1937.

ACH-20: Singeç bridge

This bridge is in the form of a deck arch with a span of 36 m on the road from Elazığ to Hozat, crossing over the Hozat river.

As described by Nafia, this bridge was the first major public infrastructure work of the Tunceli region made by the Republican Government.[65] With the completion of this project, it was possible to travel from Elazığ to Hozat in 2 hours (Figure 5.55).[66]

Singeç bridge, together with the Pertek and Gülüşkür Bridges on the Fırat River, was tendered to Aral Construction Company with a contract dated 30 March 1937 and all completed in record time.

Atatürk officially opened the Singeç bridge on 17 November 1937, also visiting Pertek bridge which was under construction, during his eastern region tour (Figure 5.56).

Figure 5.55 Singeç bridge ready for opening with the ceremonial arch erected (SALT Research, Photograph and Postcard Archive).

Figure 5.56 Left: opening of the bridge by cutting the ribbon. Right: crossing the bridge for the first time from Nafia Magazine dated 1937 (İBB Atatürk Library).

Atatürk was accompanied by the Prime Minister Celal Bayar, Deputy of Internal Affairs Şükrü Kaya, Minister of Public Works Ali Çetinkaya, General Kazım Orbay, General Inspector Lieutenant Abdullah Alpdoğan and other officials.

This visit was reported in many newspapers.[67]

On the way, crossing the old bridge[68] on Muratsuyu river, then from the area in front of the old Pertek castle, the new concrete bridge which was constructed on Hozat river was reached. The inauguration event of this bridge, which was a "high achievement of Türk Technology", was held as the ribbon was cut personally by Atatürk. Upon the information that the old name of this bridge was Soyungeç or Sungeç, Atatürk found it appropriate to call it Singeç, as it was the easiest way of pronouncing the name in our language. On the way back, the Pertek Bridge which was under construction with 100 m length over Muratsuyu was visited.

After hearing the report given by the experts about the value and importance of the Pertek bridge in terms of technical, financial and social aspects, the group headed to Pertek town centre. On the road from the town centre to the People's House (Halkevi), a large crowd cheered the great leader.

Singeç bridge is nowadays submerged in the Keban dam lake.

ACH-21: Göksu (İsaköy) bridge

This bridge is in Şile, a district of İstanbul, and crosses over the Göksu river. It is a three-hinged bridge. The approximate total length is 50 m with 5 m width (Figure 5.57).

A letter[69] in the archive informs that the price had to be increased because the abutment on the Şile side had to account for an unexpected piling, the reason being the encountered ground conditions were worse than the initial design.

It was constructed by STFA and completed in 1938.

This bridge was originally designed as a 28 m span without hinges; however, since the ground capacity in one of the abutments was found to be weaker, the structural system was revised to a 29 m span with three hinges.

The width of the road is 5.4 m with a total of 6 m between parapets.[70]

ACH-22: Sansa (Km 389) bridge

Sansa bridge is a three-span arch bridge constructed during the railway line project of Erzurum-Erzincan by SAFERHA. Its spans are 35, 20, and 20 m in the order from Erzurum to Erzincan.

The bridge is not located on the railway line route, but probably constructed to provide access over the Karasu River, which is also called the Western Fırat (Euphrates) for its great length in eastern Turkey, being one of the two tributary sources of the Fırat.

This bridge was completed in 1939. It is recorded in KGM documents, presumably because it was included in the road network later. Nowadays the bridge is abandoned with a new one built alongside serving the road network (Figure 5.58).

Figure 5.57 **During the construction of İsaköy bridge (FATEV archive).**

Figure 5.58 Sansa bridge (KGM album 1988).

Figure 5.59 Yukarıkale (Köprübaşı) bridge (author).

ACH-23 and ACH-24: Yukarıkale (Köprübaşı) and Aşağıkale bridges

These deck arch bridges with single spans are in Sivas Province and cross over Kelkit River on the Sivas-Ordu road.

Yukarıkale bridge, also called Köprübaşı, has a total length of 50 m and connects Sivas to Ordu (Figure 5.59).

Aşağıkale bridge is 65 m long and connects Sivas to Reşadiye (Figure 5.60).

Figure 5.60 Aşağıkale bridge (author).

The contractor for both bridges was Muhtar Arbatlı and Partner. Both bridges were completed in 1939.

ACH-25: KM 441 bridge

A bridge tuned with violin wire!

This single-span structure of 45 m is a railway bridge on the Erzurum- Erzincan railway line at km 441, built between 1937 and 1939. It was constructed by SAFERHA who were sub-contracted to Simeryol. The bridge was designed by Feyzi Akkaya, and Sezai Türkeş was the chief engineer responsible for construction.

During the start of the construction of this bridge, it was realized that the diagonal members of the scaffolding were missing. Therefore, the missing elements were ordered, and the scaffolding assembly started. When the scaffolding was completed, the addition of the supplied diagonals was forgotten. Thus, during the concreting of the arch, even though the concreting was made in stages, the formwork shifted laterally.

Since the diagonals of the steel scaffolding were not connected to their places, the concreted arch shifted transversely 70 cm, forming a large "S" shape on the plan view.

Then, Akkaya, who was already in Sivas, was called by a telegram to come to the site with his books. When Akkaya arrived at site, they (Sezai Türkeş, Kubas Ferit and Kopil Mitat) sarcastically told him that they brought him so far for nothing as they had already had solved this issue (not explained in the book). Akkaya insisted on visiting the bridge anyway. But, when they all arrived together at the bridge site and the engineering discussion started, Akkaya started trying to convince them of his own [better] solution, whilst also trying to be diplomatically positive towards their "solution".

He explained[71]: "...This arch can be straightened... It will be pulled from both ends towards the left and right, and forcing the bridge converge towards its axis, the diagonals would be mantled starting from the springing..."

Then, as they started arguing about his idea and Akkaya tried to defend his argument, they told him:

if you really believe in this solution then you are welcome to fix it

Suddenly, they jumped in the car leaving Akkaya on-site, slammed the car doors, and hit the gas, accelerating away. Akkaya ran behind and clung onto the windows; however, they kicked his fingers and he lost his grip. A little further away they threw out his suitcase.

Akkaya turned to Mitat, who was scared, but with a smile on his face, asked:

You didn't know, did you?

It took about a month to restore the arch back to its axis. Akkaya was ready to leave:

I invited the "conspiracy delegation" to 441, and I waved to them holding my suitcase.
-With an error of 2.5cm, your arch is ready for the concreting, commander. I am leaving!

However, they did not let him go, until the concreting was finalised. After a day of preparation, it also took another 2 days to pour the concrete of the remaining six blocks.

I tied the arch scaffolding with wire ropes, like securing a crazy [person]! to the rocks with 5ton pullers from 12 different places, and I used two 20ton capacity pulley blocks.
I could not rely on the steel truss scaffolding that had come off its axis. Therefore, I tied the strings parallel to the critical elements of the scaffolding, and I tuned them to the sound of "la" with the mouth harmonica, keeping them under control.[72] During the two days of concreting operations I was not able to get away from under the arch. My days passed by climbing like a spider and strumming these wires, examining the record of movements of the arch, brought by Kadri Veziroğlu every half hour, releasing the necessary connections, stretching the necessary ones, and yelling at the concreters above me telling them, which block should be concreted...

A multi-talented engineer indeed, it seems that Akkaya had a good ear for music and courage, demonstrated by staying under the arch as well as his engineering skills.

ACH-26: Akçay bridge

The Akçay Bridge is in Samsun province, on the Samsun to Ordu road. The structure, dated 1939 in the KGM records, was built by SAHA. The bridge is one of the rare examples of half-through bridges in Turkey. All the other half-through bridges have spans over 60 m, but, the Akçay bridge has 30 m spans as an exception for this type.

Figure 5.61 Akçay bridge from eastern end (author).

Figure 5.62 Akçay bridge pier detail (author).

There are five expansion joints in total, two at the mid-span, one of them at the pier and the other two at the abutments. Although it is not clear why the bridge had so many expansion joints, it might have been designed as a precaution against possible ground subsidence (Figure 5.61).

The bridge was originally built with three spans; however, the eastern span collapsed during a flood in 1965. The approach was extended from the eastern abutment to the pier, which became the new abutment. Therefore, the current eastern abutment of the bridge is the original pier (Figure 5.62).

Figure 5.63 **Akçay bridge from the south upstream (author).**

The bridge shows signs of deterioration, especially the hangers that transfer the deck load to the arch requiring urgent repair. During a visit in August 2017, the handrails were found to be badly deteriorated, but the bridge was still open for the local traffic. On an additional visit in 2019, the bridge handrails had been completely removed except for one section, presumably retained for interpretive purposes, and the bridge was in apparent preparation for a complete refurbishment (Figure 5.63).[73]

KGM is reportedly planning to restore the bridge.

ACH-27: Pertek bridge

Pertek Bridge had the second-longest span in Turkey with 106.9 m, crossing over the Murat River. It replaced the 216 m long timber bridge, which was 5 km upstream. Pertek bridge played an important role for local engineers, as it was built by a local engineering company (Figure 5.64).

Two bridge structures, Pertek and Gülüşkür (BWS-19), were tendered to a company named Aral Construction with a contract dated 30 March 1937, being completed ahead of schedule in 2 years. The company used highly advanced methods and technology, with machinery that was being used in the construction of similar complex structures around the world. These were cranes that moved horizontally on a cable and load carts that moved on rails mounted on the scaffold, for concrete casting. In conditions where the most basic concrete mixers were difficult to find, these vehicles were advanced technology.

Figure 5.64 **Pertek bridge (author's collection).**

Figure 5.65 **Newspaper article announcing the opening of the opening Pertek bridge.**

The bridge arch rise was 18 m with a total length of 133 m. The width was 5.85 m including 55 cm sidewalks on both sides.

The design Engineer was Emil Mörsch as shown in the project document (Figure 5.65).[74]

At the memorial meeting of Akkaya in 2005, a memory was recorded by an anonymous engineer[75]:

... around Elazığ, in the place we call Dersim, there was a 105 m span arch bridge that was still under construction at that time ... (Emil) Mörsch did the design of the project...Everyone knows Mörsch's book very well, especially in our generation...Akkaya checked the design and discussed the concept on behalf of KGM...

Davut Parker[76] worked in the bridge construction as the representation engineer of Aral Construction. The interview notes of Kemal Özcan Davaz,[77] with the Parker family (his wife Madame Tamara and daughter Madam Eteri) in their house[78] in Suadiye on 16 July 2005, are as below:

> ...Madam Eteri, also calls herself a child of the Republic as she was born in İstanbul in 1923, is healthy, straight as her mother and her memory and speech remains extraordinary. Her father, David Parker was an Engineer Class General in Tsar Nicholas' reign...
> ... Her father David Parker started to work as a contractor in the years (1927 to 1928) and built many roads and bridges in Anatolia. Meanwhile, she proudly mentioned this bridge. Her father planned and completed the construction working as a contractor in Elazığ, and showed her mother's photographs taken at the ceremony with Marshal Fevzi Çakmak and Nafia deputy Tahsin Özalp during the opening.

The bridge was on the Elazığ-Tunceli road and nowadays it is submerged in the Keban dam lake.

A letter dated 06 of June 1938 found in archives [79] is written by the Safety Division of the Interior Ministry to the Prime Minister. This letter informs that the formwork of the bridge span was completed and the military inspection group walked for the first time across the formwork from Elazığ bank to Tunceli bank. A newspaper article[80] dated 5 July 1939 announced that the bridge was soon to be opened (Figure 5.66).

Figure 5.66 Pertek bridge during construction (SALT research, photograph and postcard archive).

This letter and some photos found on the internet show that the construction of this bridge was undertaken under the direction of the military.

ACH-28 and ACH-29: Gençlik-1 and Gençlik-2 bridges

Gençlik-1 bridge (Youth bridge) was located in the city park of the new capital Ankara. This pedestrian bridge had an atypical design and it was part of the landscape. Its construction is recorded in 1939 in the KGM records. Photos of the construction of this bridge can be seen in magazines of 1939. The Park was opened on 19 May 1943 (Figure 5.67).

There were two arch bridges when the Park was completed; however, the shorter one, Gençlik-2, was not as well recognized as the longer one, which became the symbol of the Park.[81]

The Park had a pioneering role in the transformation of the socio-cultural life in the new capital in the 1940s with the facilities including a small artificial lake, an open-air theatre, sports areas, tea gardens and restaurants for leisure and recreational activities. Hermann Jansen planned the park in 1933, but the final design was completed by the French architect Theo Leveau, with some alterations (Figure 5.68).[82]

The lake was created for landscaping but was also deep enough for boats and canoeing and fed by the İncesu creek. The circulation of pedestrians within the park was planned to be provided by two identical arch bridges with 40 m span in 1933 as proposed by Jansen. On later revision, one of the bridges was shorter than the other in the plans modified by Theo Leveau in 1935.

Figure 5.67 Gençlik-1 bridge (author's collection).

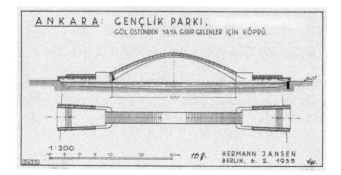

Figure 5.68 Proposed Gençlik bridge from Hermann Jansen Plan (Architektur museum TU Berlin, Inv. No.: 22890).

Figure 5.69 Gençlik-I bridge from 1952 dated 100 lira Banknote.

The construction of the bridge was awarded to "Eng. J. Acıman"[83] as shown in the KGM records. Jacques Aggiman (1892–1942) was an Ottoman born Canadian citizen and engineer, mostly engaged in the construction of buildings for embassies and worked as a contractor in Ankara between 1925 and 1942.

The bridge also became one of the well-known symbols of the park and of Ankara; it was used for postcards and pictures of Ankara, as well as in a banknote of the Turkish Lira.

The bridge is very slender and elegant with a fine appearance. The structural elements of the bridge are carefully detailed. The bridge deck is a continuous thin element from one end to the other. The arch, which constitutes two ribs, is shaped like a crescent. The supports of the arch and column at the springing are well-proportioned where they share the same seat. The deck rests on approach columns beyond the arch and descends with a gentle slope. The approach columns are tapered with a smaller base. The arch ribs are interconnected with bracing members placed at regular intervals (Figure 5.69).

The resemblance of the bridge to the Leonardo da Vinci bridge proposed for Galata in 1495 is noticeable. Especially the arch and deck profiles united in the middle and with a deck continuing beyond the ends of the arch.

The crescent shape of the arch implies hinges at the springing. The structural design is unique in this sense; the arch can also be interpreted as an inclined leg of a frame bridge with a haunch where the deck and arch merge. The column at the springing is presumably structurally connected with the leg or arch as perceived in the photos, providing a triangulation effect to transmit the forces beyond.

The span for Gençlik-1 was approximately 35 m, the length having been reduced by the platforms extending from the banks of the lake. This bridge was demolished and replaced in the 1980s as a part of a Park renewal project. The profile of the new bridge remains similar, but the architecture has been modified.

Gençlik-2 bridge, which has a shorter span of approximately 15 m, connects the former coffee house, currently being used as a wedding hall, to the park itself. The current structure in place at this location is very similar to the original bridge, but the deck has steps at the entrances. It is not clear if the smaller bridge is a replacement or a result of refurbishment works (Figure 5.70).

ACH-30: Pisyar Bridge

This bridge is on the Batman-Kozluk-Bitlis road crossing over the Garzan river. It is a single span of 45 m with three hinges.[84]

Pisyar and Anbarçayı bridge were tendered together[85] and both were awarded to SAHA on 23 September 1937.

As we learn from Akkaya[86]; Türkeş worked on the Pisyar bridge on the Silvan to Hazro Road and handed it over to Bodos Kemal after completing half of the structure. Akkaya also visited the bridge himself, while he was working on the construction of the bridges of the Diyarbakır to Kurtalan Railway line. He visited the construction site with Raşit as they had a 24-hour leave from their work. After their visit, the truck did not come to pick them up and they were forced to walk for 2 days from Pisyar bridge to the rail line site.

Figure 5.70 **Gençlik-2 bridge (author's collection).**

The bridge was partially damaged by a flood in 2004, when the spandrel column failed, resulting in part of the deck also collapsing. Since then it is used by locals as a jetty for fishing.[87]

ACH-31, ACH-32 and ACH-33: Alikaya, Suçatı and Tekir bridges

These three bridges were built on the Elbistan-Maraş road. Alikaya and Suçatı crossed over the Ceyhan River, and Tekir bridge was over Tekir creek. The only documentation found is from the records in STFA archives and Nafia Magazine.[88] The three bridges have the same width of 5.4 m for road and 6 m between parapets.

Alikaya bridge had a 45 m span and it was a three-hinged bridge. It is now submerged in the dam lake of the Menzelet dam (Figure 5.71).

Suçatı bridge has a 35 m span and is also designed with three hinges. It is now submerged in the dam lake of the Menzelet dam (Figures 5.72 and 5.73).[89]

Tekir bridge has a 21.9 m span arch with a continuous crown. It is currently existing and serving for local traffic to Suçatı village (Figure 5.74).[90]

Akkaya[91] mentions the bridges as:

> ...Sezai was busy 'loosening the knot' on the four bridges named Alikaya-Suçatı-Tekir-Eloğlu on the Pınarbaşı-Maraş road, which the company (SAHA) was contracted. His trusted worker Salih Güzey, covered with blood in his chest, diving in the river with his shirt, bringing the lumbers from the river for the scaffolding. Master (Dimitri) Kuru and Master Rıfat...were racing to keep up with Sezai ...

Three tender advertisements dated 16 September 1936, 25 March 1937 and 30 April 1937[92] are found in newspaper records, which is an indication that the bridge had some challenges before its start.

Figure 5.71 Alikaya bridge (Photograph: Prof. Durmuş Öztürk).

Figure 5.72 Suçatı bridge (Photograph: Prof. Durmuş Öztürk).

Figure 5.73 Suçatı bridge (Photograph: Prof. Durmuş Öztürk).

ACH-34: Goat (km 386) bridge

Although the bridge was built as a road bridge to provide the connection for the Sivas to Erzurum Railway line, it was soon abandoned. Therefore, it does not appear on the road network, nor in the KGM records.

This half-through bridge has a span of 68 m and was built by SAHA, the construction started in 1939 (Figure 5.75).

Akkaya mentioned the bridge[93]:

We quickly completed Erzincan, then headed towards Erzurum and the Sansa Bridge [Km 389] next to the Musa Rock, which we built and left 20 years ago, and

Figure 5.74 Tekir bridge (Photograph: Yusuf Köleli from https://marasavucumda.com).

Figure 5.75 Goat (Km 386) bridge scaffolding preparation on the ground (STFA archive).

Km 386 Bridge, which we built for the road junction, started to line up one by one like old friends. Km 386 Bridge, which was abandoned because the junction was abandoned, was called the "Goat bridge" by the villagers!...

This bridge was completed in 1940.

ACH-35: Dicle (Eğil) bridge

The 1952 dated single-span arch bridge crosses over Dicle river with 80 m span length and a total length of 140 m. The width of the bridge is 5.9 m. The structure is designed as fixed ended support without hinges. Bridge approach spans are of beam type supported on slender columns piers. These approach columns are detailed in a similar fashion as the arch spandrel columns (Figure 5.76).

The bridge will be submerged in the lake of Kralkızı Dam.

ACH-36: Mameki bridge

Dated in 1952, this bridge is on the first kilometre of the Tunceli to Elazığ Road over the Munzur river. The total length is 116 m with a span of 72 m (Figure 5.77).

It appears likely that this bridge shares a similar design as the Irmak bridge (Figure 5.78).

Figure 5.76 Dicle bridge (KGM album 1988).

Figure 5.77 Mameki bridge (KGM album 1969).

Figure 5.78 Mameki bridge (KGM album 1988).

ACH-37: Irmak bridge

This 1954 dated bridge is on the Ankara-Kırıkkale road. Its total length is 116 m with a span of 72 m. It is a three-hinged bridge (Figures 5.79–5.81).

Figure 5.79 Irmak bridge (KGM album 1988).

Figure 5.80 Irmak bridge (author).

ACH-38 and ACH-39: Mutu and Sansa bridges

These are two half-through single-span bridges both crossing over the Karasu River. Mutu is on the Erzincan-Tunceli road with a span of 60 m. Sansa is on the Erzincan-Erzurum road with 80 m span. Both structures demonstrate a majestic appearance

Figure 5.81 Irmak bridge hinge at springing (author).

Figure 5.82 Mutu bridge (KGM album 1988).

with similar designs in different span lengths. Mutu bridge has 12 hangers while Sansa bridge has 18 hangers (Figures 5.82–5.84).

Mutu and Sansa bridges were both awarded to SAFERHA with a contract dated 22 November 1950. They were completed by SAHA in 1954.

Figure 5.83 Sansa bridge (KGM album 1988).

Figure 5.84 Sansa bridge (KGM album 1969).

ACH-40: Kütür bridge

This bridge opened to traffic in 1954 and was tendered together with Mutu and Sansa bridges. It is in Erzincan province on the Erzincan-Erzurum road crossing over the Karasu River. It is located at a significant crossing, where the abandoned masonry bridge and operational railway bridge are currently standing. Kütür bridge is well hidden behind them when observed from the current highway bridge (Figure 5.85).

It is a single-span bridge with a 62 m span and a total length of 107.5 m.

ACH-41: Kurtuluş bridge

This 20 m single-span bridge is in the Ordu province on the Black Sea region along the Ordu to Fatsa Coastal Road. The road network between Ordu to Fatsa was constructed crossing through the mountainous region; this road was constructed by locals in 1932 (Figure 5.86).

Figure 5.85 Kütür bridge (KGM album 1969).

Figure 5.86 Kurtuluş bridge from the south upstream side (author).

Figure 5.87 Kurtuluş bridge from the south upstream side (author).

Figure 5.88 Soğanlı bridge (KGM album 1988).

The road network relocated to the coastline which runs parallel to the Black sea coast. This Bridge was finished in 1955 as a part of the Coastal Road Project.[94] The bridge has a single hinge at the crown (Figure 5.87).

ACH-42: Soğanlı bridge

This bridge spanned Soğanlı creek on the Karabük to Gerede road. It is dated 1955 and it is made of a single span of 68 m (Figure 5.88).

ACH-43: Çayırhan bridge

Çayırhan bridge is in the Ankara province on the road from Nallıhan to Çayırhan. It is constructed with the Sarıyar dam project, which is on the Sakarya River and was built between 1951 and 1955. It was open to traffic in 1955 (Figure 5.89).

The bridge was required to provide transportation when the water level in the dam lake rises (Figure 5.90).

This 42 m span bridge has a rise of 24 m and a width of 9.4 m. The rise/span ratio of the bridge is around 1/1.75.

Figure 5.89 Çayırhan bridge (KGM album 1988).

Figure 5.90 Çayırhan bridge and author (M.Arch. Sevil Işıl Ören).

Figure 5.91 Malabadi bridge (KGM album 1988).

ACH-44: Malabadi bridge

Malabadi bridge stands in the shadows of the well-recognized magnificent historical Malabadi bridge constructed between 1145 and 1154 by the Artuqids. This stone
bridge has a remarkable clear span of 38.6 m and is considered as an engineering feat
of its days. It has unique features such as accommodation areas located inside the
spandrel volume and accessed by stairs.

The concrete deck arch bridge opened in 1955 on the upstream side of the pre-
existing historical bridge with a 56.5 m span on the Diyarbakır-Siirt road (Figure 5.91).

KGM Director, Orhan Büyükalp[95] provided a full account of the construction of
the bridge:[96]

> … In 1951, the bridge was tendered to the contractor. Since his possibilities were
> limited, he could start the work with a delay in 1954, putting an effort, which can
> be described as "'taking the bit in his teeth [making a great effort]".

The contractor, then, proceeded with the work, until the middle third width strip of the
arch was poured, and only 24 hours later, a terrific flood came and took away the entire
scaffold; in consequence, the planks and poles drifted in the flood water (Figure 5.92).

At that time, Büyükalp was the construction chief in the region and construction
was entrusted to his department.

> I did not have the slightest experience in building a bridge like this with a 56 m
> span reinforced concrete arch…But in those years, all of us – including me - had
> a very necessary and important motivation and excitement to accomplish a work.
> The enthusiasm of belonging to "Karayolculuk/ Roadbuilder" and the commit-
> ment and determination to get the job done without minding if it was easy or hard,
> embracing it with four hands and finish it at any cost.

Figure 5.92 Malabadi bridge the collapsed scaffolding of the bridge (SALT Research, Ali Saim Ülgen Archive).

Büyükalp was anxious and fearful, he was especially worried that such a flood may come again during the remaining of the construction. Then he thought of making the scaffolding safe against flood and decided to construct the new scaffold from steel.

> I shared my plan with my friends/colleagues in the bridges department of KGM. They said: "First of all, there are cracks [in the ground] at the springing, secondly, we cannot rely on the welded steel system in the scaffold, so this plan does not sound achievable". I kept on visiting them the following days.
>   In the end, they concluded: "If you believe and makes sense for you, go and do it as you insist. You will be returned a dry "good job" or "thank you", if you succeed, but if a disaster happens and scaffold fails, do not refer at us, we neither talked to you and we do not know about it and don't expect for the slightest protection and support from us".

Büyükalp made a decision to carry on. He roughly calculated the quantities required for a steel scaffold and found that 20 tons of L shaped steel profile and 20,000 electrodes would be needed. However, as he would not be known by anybody in the steel market in Ankara, and because the General Directorate did not accept the idea of welded steel scaffolding, they would not assist in the procurement of materials either. Then, he contacted a well-known contractor Muammer Kıraner,[97] who immediately ordered the needed material, on his account.

Büyükalp did not lose any time:

...We loaded the material and departed to Diyarbakır. At that time, the Birecik bridge was not built on the Fırat (Euphrates) River. Trucks and passengers had to cross the Fırat River in boats.

After their arrival in Diyarbakır, they prepared the scaffold in the workshop. In the meantime, the crew strengthened the cracks in the springing level at the bridge site. Then, the scaffold truss was transported to the bridge site. The scaffolding was brought under the existing one-third width of the concrete arch on wooden barges. The Scaffold truss was tied at both ends to the pulley ropes, which were mounted on the top of the arch, and truss pieces were raised until they came in contact with the arch intrados, then the connections were fixed through welding, until all the pieces formed the arch.

The scaffolding was completed in a short duration, of about 15 days.

Malabadi Bridge with 56 m concrete arch span and 100 m long total length was completed in 3.5 months; it was opened to service in March 1955.

Büyükalp also presented an album showing the various stages of construction. He received a letter of appreciation as transcribed below:

Orhan Büyükalp should be rewarded with a commendation for completing the reinforced concrete Malabadi Bridge, which has technical significance, with a 56 m arch span, in a short period of three and a half months

When the concrete arch bridge was completed in 1955, the old bridge that had served for 801 years, was indeed still in use for traffic and also serving for accommodation, as we learn from Büyükalp:

... While I was working on the Silvan-Kozluk road as a site manager, in 1951, we had sheltered in the old bridge with our workers for a whole season. In the old bridge, there are two large rooms which can be reached by steep stairs, on the right and left sides of the crown. We allocated one of the rooms as a dormitory for workers and the other as the kitchen. The windows in these rooms facilitated ventilation and lighting.

A photograph clearly showing the image of the old and new Malabadi Bridges was printed on a series of commemorative stamps prepared for the International Roads Congress held in İstanbul in 1955.

Malabadi concrete arch bridge is now itself a heritage for both representing a certain era and its engineering features. Nowadays, the two bridges together create a rather interesting view providing a contrasting background to each other.

ACH-45 and ACH-46: Dokuzdolambaç bridges

Two virtually identical bridges called, Dokuzdolambaç 1 and Dokuzdolambaç 2 are crossing over Kelkit river on Tokat-Reşadiye road. The first bridge was finished in 1955 and the second in the following year.

Figure 5.93 Dokuzdolambaç-1 bridge (KGM album 1988).

Figure 5.94 Dokuzdolambaç-1 bridges (author).

Each of them has a single span of 60 m and is 9.2 m wide (Figures 5.93–5.95).

ACH-47: Dergalip bridge

The 1956 dated bridge is a 60 m single-span arch with a total length of 98 m. It is similar in design to Hasankeyf Bridge (Figure 5.96).

Nowadays it is mostly submerged in the dam lake of Ilısu.

Figure 5.95 Dokuzdolambaç-2 bridge (KGM album 1988).

Figure 5.96 Dergalip bridge (KGM album 1988).

ACH-48: Birecik bridge

Birecik bridge was Turkey's longest concrete road bridge with approximately 720 m total length at the time of its completion on 14 March 1956.[98] The Bridge, located in the Birecik district of Şanlıurfa province and, crossing over the Fırat (Euphrates), lands to the Nizip district of the province Gaziantep on the opposite bank of the river.

This bridge can be divided into two parts structurally. The first part on the Birecik side has an arch deck system with spandrel columns. This part consists of five arches,

Figure 5.97 **Birecik bridge (KGM album 1988).**

each of them with a span length of 57 m. The total deck length of the arch section is approximately 300 m. The second part of the bridge on the Gaziantep side has a beam superstructure of 14 spans, each 26 m long, totalling another 364 m. Between these two parts, there is a pier/abutment of 15 m and on the eastern end another 20 m approach beam span connected to the bridge with another 15 m pier/abutment (these are all approximate dimensions taken from old sketches). The total width of the bridge is 11 m which includes 1.5 m pedestrian sidewalks on either side (Figure 5.97).

The river passes under the arched portions on Birecik side, the approach spans across the flood plain on the Gaziantep side.

The bridge has very frequent joints; in the arch part: there are four joints in each span and in the beam part, there are two joints in every third span. Joints are provided so frequently to compensate for the differential movements from the poor ground conditions.

The first studies for the Birecik Bridge were made in 1912. Mr Nadir, who was Chief Engineer of the Aleppo Province sent a report stating the importance of the bridge to be built in Birecik, along with other projects that he had prepared, to the government.[99] However, the Balkan War, and then WWI, made most such projects virtually impossible.

In 1933, in the tenth year of the new Republic, the Birecik bridge project was given further consideration. However, it did not proceed due to insufficient funds. In those years, the continuous interruption of transportation, the trucks waiting in long queues in Birecik and the lives lost in the flow of water while the raft passed from one side to the other, reinstated the need for a new bridge to be built here.

The survey and ground investigation works of the bridge were undertaken by Feyzi Akkaya. KGM was preparing the projects and the bill of quantities. The necessary allocation for construction was included in the 1951 budget. The tender was announced on 15 June 1951. Fifteen companies showed an interest in the project, but only four of them participated in the tender. The contract was awarded to the Amaç Turkish Joint Stock Company (Amaç Türk Anonim Şirketi), and a contract was signed on 31 July 1951.

Figure 5.98 Birecik bridge (KGM calendar album 1974).

According to the contract, the duration of the work was scheduled to be undertaken in 24 months. However, an extension period of 26 months was signed for the completion due to unexpected floods caused by various reasons and especially by the abnormal flow in the Fırat river after the winter of 1952 (Figure 5.98).

There is a related article[100] in Akis magazine in 1956 found at İnönü Foundation Archive. This article provides detailed information regarding the bridge, especially its economical justifications:

> Long ago, the president of the time had to cross the Fırat (Euphrates) with a temporary pontoon bridge formed by rafts lined up next to each other. That day, a bridge was promised to the people of Birecik. But they would have to wait...
>
> Today, the Birecik Bridge, which connects the east and west of Turkey, is in a way, cheap: 4400 tons of cement, 921 tons of steel were used, 17000 m³ concrete was made and 250 workers laboured on the construction. The bridge was completed in four seasons. Considering the economic importance of the Birecik Bridge, the time to be saved and the unnecessary expenses to be prevented, the bridge is "cheap". Once Birecik Bridge is open, its traffic will undoubtedly increase with an estimated growth of 50% every following year.
>
> According to the statistics in 1955, on average, 530 tons of various cargo arrive in Birecik, and 225 tons need to cross to the opposite banks on a daily basis. The average number of travellers per day was 1459 and 724 were cross passengers. 233 various vehicles arrive, 85 of which were using the rafts to cross. Now, these numbers are expected to rise rapidly. Because now there is no hassle of unloading the 10ton truck, then crossing the river and then load it again, there is no danger of getting lost in the waters of the Fırat.
>
> The unloading and loading of the goods and raft money are added as expenditures to the transport. A vehicle would wait half a day before the bridge was built.

From now on, trucks and cars will continue on their way without stopping. Apart from the economic benefits of the bridge, the bridge will be built "free of charge", not "cheap", within 2.5 to 3 years, if only time saving considered.

Since the producer will no longer worry about his goods being rotten on the roads or sinking into the waters of the Fırat, there will also be a development in economic activity. The crops and products of the Fırat Valley, which have not reached the domestic markets until now, will be able to be distributed within the country within one day.

The contractor went through four site chief engineers within roughly two years. These site engineers were Ertuğrul Barla, Adnan Arbatlı, Kadri Çile and Suavi Atasagun in that order. The third one, Kadri Çile, was killed in 1953 by a labourer who was fired while on duty at the construction site. His tomb was placed nearby the bridge.[101] The Control engineer was Mustafa Tanrıkulu, who lasted the entire project from the beginning to the end.

KGM Engineer Halim Ağaoğlu, a colleague of Tanrıkulu, visited the construction and provides a detailed account of the works: [102]

The most difficult phase of the construction was the lowering of the foundations in the water to a depth of 8 m and placing them on the rock layer below. For this work, 12 m long Larssen imported from Germany were used. A 300 m long service bridge was built on wooden piles for the construction of arch spans... This service bridge could not withstand the flow, it was collapsed and many construction machineries were submerged in the great flood caused by the rise of the waters of the Fırat. Unfortunately, a construction foreman got lost in the turbid waters of the Fırat and was never found again...

Akkaya[103] mentions the bridge as follows:

... we were preparing an offer for Birecik Bridge on Fırat. We also made investigations for this bridge and understood the ground conditions well and we also knew what to do. There was also lumber from other bridges ready for this job. Despite this, we were able to offer only a small discount, however the newly established "Amaç" company made a very cheap offer and won this job from our hands...

A year later, ... I stopped by the bridge on my way. The silence and inactivity on their construction sites immediately caught my eye. After a year or two, "Amaç" was able to finish this bridge, but at the same time, it ruined itself too...

Birecik bridge image has been printed on a 1959 industrial series postage stamp.

As of today the bridge still stands and has proven its feasibility as a main connection in the region.

ACH-49: Ceyhan bridge

This 1958 dated bridge was on the Maraş to Göksun road. It is now submerged under the Kılavuzlu dam lake on the Ceyhan River. The single-span bridge had 77 m total length and a 60 m deck arch span (Figure 5.99).

Figure 5.99 Ceyhan bridge (KGM album 1969).

Figure 5.100 Tabakhane (Balya) bridge (Ertuğrul Ortaç/www.ortac.net).

ACH-50: Tabakhane (Balya) bridge

Tabakhane bridge is still in service on the road connecting Balikesir to Çanakkale. This single-span bridge is unique with its four ribs interconnected with transverse beams (Figure 5.100).

The original guardrails and the entrance bollards are still in place. This Bridge is dated 1959 in the KGM records; however, the features of the bridge, guardrails and bollards suggest that it may have been constructed at an earlier date.

Figure 5.101 Sirya bridge (KGM album 1988).

Figure 5.102 Maden bridge (KGM album 1988).

ACH-51: Sirya (Zeytinlik) bridge

This bridge is in Artvin province on the road from Artvin to Erzurum, crossing over the Çoruh River. It was built in 1960 and its name was taken from the nearby settlement Sirya also called Zeytinlik (Figure 5.101).

According to the KGM records, it has a span of 67.5 m, with a total length of 88 m. The bridge will be submerged in the Deriner dam (or already is).

ACH-52: Maden bridge

Maden bridge is dated 1962 and is on the Elazığ to Maden Road crossing over the Dicle River. It has a 40 m single span, with a total length of 120 m. The width of the bridge is 9.7 m (Figure 5.102).

Figure 5.103 Hasankeyf bridge (KGM album 1988).

This bridge is different with its notable high rise/span ratio sharing this feature together with the Çayırhan Bridge.

ACH-53: Hasankeyf bridge

This three-span bridge was finished in 1964 as a deck arch bridge crossing over Dicle (Tigris) River with spans of 60, 50 and 40 m in the order from Batman to Midyat. The bridge had two ribs, at the edges of the deck, interconnected by beams at the same intervals with spandrel columns (Figure 5.103).

There was an ancient stone bridge which was also called Hasankeyf bridge on the upstream side of this structure. From the ancient bridge there are remains of three piers and a small arch from what could have been an amazing and fantastic bridge. This was one of the well-recognized heritage places in Turkey (Figure 5.104).

Hasankeyf was the capital of the Artuqids between 1101 and 1232. The Ancient Bridge was built by locals between approximately 1147 and 1172, with a main span of 45 m, tall piers and an open channel crossing around the piers.[104]

Both bridges are now under the lakes of Ilısu Dam.

ACH-54: Cizre bridge

Cizre bridge is dated 1967 and is on the Cizre- Şırnak road crossing over the Tigris river. It has three spans of 60 m with a total length of 192 m. The width of the bridge is 11.50 m. The arch has two wide ribs and its columns support the deck over the arch.

As learnt from the memoirs of Ağaoğlu, the structure was tendered on 16 March 1960[105] to a contractor; however, the construction could not proceed and twice the excavation was sabotaged. Hence, the tender had to be liquidated, and KGM resumed the construction works with its own resources. The military was of assistance in order to complete the bridge (Figure 5.105).

Ağaoğlu's[106] observations on his second visit to bridge site in 1966:

Figure 5.104 **Hasankeyf bridge (author).**

Figure 5.105 **Cizre bridge (KGM album 1988).**

... The destruction of the steel sheet piling plates driven for the foundation excavation could not have been caused by floods. This destruction had happened
twice and with a strong possibility, it was done by dynamite. Barzani militants
might have believed that if the bridge is constructed, the illegal food trade could
be in danger... In Cizre, we immediately took all the precautions to anticipate this.

All workers were changed and the construction site was protected by military force. Selçuk Hayırlıoğlu, who is my dear friend, works in the General Directorate as a bridge construction expert, was transferred to manage this job. The bridge was completed within one year and opened to traffic.

NOTES

1. MacCurdy, E. (1955) *The Notebooks of Leonardo da Vinci*. George Braziller. New York.
2. Benaim, R. (2008) *The Design of Prestressed Concrete Bridges Concepts and Principles*. CRC Press. UK. Chapter 7.
3. Troyano, L.F. (2003) *Bridge Engineering: A Global Perspective*. Thomas Telford. London. Page: 274.
4. Dufour, F.O. (1908) *Cyclopaedia of Civil Engineering*. American Technical Society. Chicago.
5. The term used for Public Works was "imar", means civilization and development excluding the social content. Term later changed to "Nafia", which can be translated as utility, to refer to public assembly at the governance level. Nafia also later changed to "Bayındırlık", means prosperous, in 1935.
6. Nafia (1933) *On Senede Türkiye Nafiası 1923–1933*. Nafia Vekaleti Neşriyatı. İstanbul.
7. **Risorgimento** is the first reinforced concrete bridge in Rome over the Tiber dated 1911. It has 100 m single span with a very low rise/span ratio using hollow box structure designed with the system of François Hennebique.
8. *Joseph Monier* (1823–1906) was a gardener, also producing gardening items: planters, pipes and tanks made with concrete and iron mesh. He patented the idea of strengthening thin concrete tubs by embedding wire mesh in the concrete in 1867. In 1873, he patented a new low rise/span ratio arch bridge system and in 1875, built the first reinforced concrete bridge.
9. Troyano (2003) Page: 189.
10. Troyano (2003) describes this design and narrates his father, Carlos Fernandez Casado, calling them "anachronic bridges" for their unique application in Spain. Page: 189. Tyrell (1911) has a dedicated chapter for these bridges, named 'solid concrete bridges' and provides a list of eight of them spanning more than 200 ft (60 m).
    Tyrrell, H. G. (1911). *History of Bridge Engineering*. Chicago. H. G. Tyrrell.
11. BOA PLK-p-3898 Ayvalık-Bergama yolu üzerine yapılacak köprünün planı. (TRM 1634, Fr.)
12. BOA PLK-p-3576 Daday Çayı üzerine inşa olunacak köprünün plan ve resmi. (EHT.)
13. *François Hennebique* (1842–1921) a mason turned contractor in Paris, experimented with concrete reinforced with iron from about 1879, and in 1892 patented his system, giving his method of calculation and typical reinforcement details for the bending of beams and slabs. He established an empire of franchises in major cities and is accounted responsible for the rapid growth of reinforced concrete construction in Europe.
    Cusack, P. (1984) François Hennebique: The Specialist Organisation and the Success of Ferro Concrete 1892–1909. *Transactions of the Newcomen Society*, LVI, 71–86.
14. Frapier, C., and Vaillant, S. (2012). The organization of the Hennebique firm in the countries of the Mediterranean Basin: Establishment and communications strategy. In Piaton, C., Godoli, E., and Peyceré, D. (Eds.), *Building Beyond the Mediterranean: Studying the Archives of European Businesses (1860–1970)*. Publications de l'Institut national d'histoire de l'art. Arles. DOI: 10.4000/books.inha.12692.
15. Similar bridge drawings for the same project can be found at BOA 2681, 4978 and 5918.
16. *Robert Maillart* (1872–1940) an engineer who revolutionized reinforced concrete with such designs as the three-hinged arch and the deck-stiffened arch. Maillart had an intuition and genius that could entirely exploit the aesthetic of concrete.
17. *Karl William Ritter* (1847–1906) was a professor of the Swiss Federal Institute of Technology Zurich, and later rector of the Polytechnic Institute of Zurich.
18. Billington, D. (1983) *The Tower and the Bridge*. Basic Books, Inc. Publishers. New York. Pages: 161 and 163.

19. İlter, İ. (1974) 51 Yılda Köprülerimiz ve Köprücülüğümüz. Karayolları Magazine. Year: 23, No: 295–296.

20. **Emil Mörsch** (1872–1950) studied civil engineering at Stuttgart TH. He worked in the Ministerial Department for Highways & Waterways, and afterwards was employed in the bridge unit of State Railways. He joined the Wayss & Freytag company in early 1901. From 1916 onwards, Mörsch worked as professor of theory of structures, reinforced concrete construction and masonry arch bridges at Stuttgart TH.

He published the first edition of his book 'Der Eisenbeton: seine Theorie und Anwendung', which underwent numerous reprints. This book set used as a standard in reinforced concrete for more than half a century. Book was well known in Turkey.

Bozdoğan, S. (2001) *Modernism and Nation Building: Turkish Architectural Culture in the Early Republic*. University of Washington Press. Seattle.

21. KGM unclassified documents Binder no: 2092, drawing prepared by Société Anonyme Turque d'Etudes et d'Entreprises Urbaines Constantinople. Found in: Örmecioğlu, H. T. (2010) Technology, Engineering, and Modernity in Turkey: The Case of Road Bridges between 1850 and 1960. Ph.D. thesis submitted to Middle East Technical University (METU). Page: 122.

22. The title of the drawing is given as: "Projeckt einer Bogenbrucke de (Irva) Deressi bei Boshane von 50 m Spannweite" (Project of an arch bridge in (Irva) Deressi near Boshane with a span of 50 m).

23. Nafia (1933)

24. Figure H. 10 Reinforced concrete bridge technology had remained almost the same until 1960. Drawings of two reinforced concrete arch bridges Akçay Bridge in 1928 (above) and Hasankeyf Bridge in 1956 (below). Source: Unclassified documents from the State National Archives-Republican Archives, KGM Fund, Binder no: 9616-1 and 9616-2. Found in: Örmecioğlu (2010) Appendix H.

25. Strassner, A. (1949) Yeni Metodlar Cilt: 2 – Kemer Ve Kemerli Köprü Statiği. ITU translated to Turkish by Berkman, A.F.

Original book: Strassner, A. (1927) *Neuere Methoden zur Statik der Rahmentragwerke und der elastischen Bogentrager*. Aufl. Ernst & Sohn. Berlin.

26. Göksu Köprüsü (1932 March 7) Sehir Haberleri. Akşam. Page: 3. Column 5.

Göksu Köprüsü (1932 Mach 8) Belediye Haberleri. Milliyet. Page: 2. Column 6.

27. Unclassified documents from the State National Archives-Republican Archives, KGM Fund, Binder no: 16041. Pages from letter from SaFerHa to Nadir Bey, the Director of the Roads and Bridges Department, on 28.8.1928. Found in: Örmecioğlu (2010) Page: 127.

28. BCA 30.18.01.02. No: 9. 11. 5. Presidential decree on construction of Kömürhan Bridge signed by Gazi Mustafa Kemal.

29. NOHAB (Nydqvist & Holm AB) was a manufacturing company based in the country of Sweden.

30. Christiani & Nielsen was established in Copenhagen in 1904 as a contractor company to build bridges, marine works, and other reinforced concrete structures.

31. NOHAB was contracted for Irmak-Filyos and Fevzipaşa-Diyarbakır line during 1927–1937.

Nydqvist & Holm, Construction Des Lignes, De Chemins De Fer Irmak – Filyos & Fevzipaşa –Diyarbekir. Göteborg, 1937, Page: 52.

32. Values corresponds to 8 MPa and 5 MPa respectively. Nafia 1933 dated standard describe normal concrete as 300 kg/cm^2 (30 MPa without safety factor).

33. Figure 4.21 Strength results of test beam with Dykerhoff-Doppel super cement. Source: Unclassified documents from the State National Archives-Republican Archives, KGM Fund, Binder no: 16041. Found in: Örmecioğlu (2010) Page: 134.

34. Pucher (1937) Zwei Eisenbetonbrucken uber den Euphrat. Der Bauingenieur v.18. Heft:5/6, 05 February 1937 Page: 67.

35. BCA, 30.10.0. no: 155. 90. 8. Presidential decree on naming the bridge signed by Gazi Mustafa Kemal.

36. Translated by Author.

37. Kann, F. (1936) Büyük Açıklıklı Bilhassa Kemerli Massif Köprülerin Projelerinin Tanziminde Nazarı Dikkate Alınacak Umumi Kaideleri Bayındırlık İşleri Dergisi (Yönetsel Kısım), Y.2, S.8, Teşrinisani 1936, Page: 42. Translated by Fuad Özdeğer.

38. Asyada Birinci Gelen Büyük Beton Köprü (1932 June 1) Son Posta Newspaper. Page: 9.
39. Memlekette Köprü Siyaseti İyi Neticeler Veriyor, Asırlardan beri yapılamayan Köprüler 3 yıl icinde nasıl Vücuda getirildi ? (1932 August 15) Son Posta Newspaper. Page: 9.
40. **Fevzi Çakmak** (1876–1950) was a Turkish field marshal (Mareşal) and politician. He served as the Chief of General Staff from 1918 and 1919 and later the Minister of War of the Ottoman Empire in 1920. He later joined the provisional Government of the Grand National Assembly and served as the Deputy Prime Minister, the Minister of National Defence and later as the Prime Minister of Turkey from 1921 to 1922.
41. **Feyzi Akkaya** (1907–2004) a prominent bridge engineer of the early Republican era. Together with **Sezai Türkeş** founded STFA Company, which was a leading engineering company in Turkey and worldwide. Their inventions and novel applications have been a vital contribution to the engineering field in Turkey. See Chapter 3 for expanded biography.
42. Akkaya, F. (1989) *Ömrümüzün Kilometre Taşları: STFA'nın Hikayesi.* Bilimsel ve Teknik Yayınları Çeviri Vakfı. İstanbul. Pages: 36–37.
43. **Kemal Hayırlıoğlu** (1891–1984) was Nafia Chief Engineer for 15 years between 1928 and 1943. He worked in various positions in Nafia until retired in 1955. See Chapter 3 for expanded biography.
44. It was a style for engineers to wear fedora style hat.
45. Adana Nafia Başmühendisliğinden (1932 October 20) Tender Announcement of the Bridge. Cumhuriyet Newspaper. Page: 5.
46. **Halit Köprücü** was a prominent bridge engineer and one of the founders of SAFERHA, which constructed the most bridges in both number and significance in Turkey. He later worked in KGM. See Chapter 3 for expanded biography.
47. Engineer Köprücü, H. (1934) Körkün Köprüsü. Design: Engineer Halit Project: Construction: Engineer Ferruh, Sadık, Halit. Arkitekt Magazine No: 1934–08 (44). Pages: 233–234.
48. **Eugene Freyssinet** (1879–1962) was a French engineer discovered the phenomenon of creep in concrete and developed the techniques of prestressed concrete, although he was not its inventor.
49. Akkaya (1989) Page: 35.
50. **Sezai Türkeş** (1908–1998) a prominent bridge engineer of the early Republican era. Together with **Feyzi Akkaya** founded STFA Company, which was a leading engineering company in Turkey and worldwide. Their inventions and novel applications have been a vital contribution to the engineering field in Turkey. See Chapter 3 for expanded biography.
51. Akkaya (1989) Page: 36. Another event that Akkaya made an early decision about the concreting the foundation of Körkün Bridge and was left idle for a while by his manager.
52. Avrupada Misli Nadir Betonarme Yeni Bir Köprümüz Daha İkmal Edildi (1934 February 15) Milliyet Newspaper. Page: 6.
53. Engineer Halit (1935) Pasur Köprüsü. Design: Nafia Vekaleti Fen Heyeti. Construction: Engineer Ferruh, Sadık, Halit and Kemal. Arkitekt Magazine No: 1935-03 (51). Pages: 65–67.
54. Akkaya (1989) Page: 35.
55. The shrinkage of the concrete is the reduction of its volume during its hardening. The main reason for this is drying, that is the loss of water once the fresh concrete starts hardening as a result of a chemical reaction generating heat. For this reason, it is a rational method to reduce the shrinkage by pouring the concrete in cold weather.
56. Akkaya (1989) Page: 54.
57. Vamık Faik (1934 April 15) Dördüncü Büyük Köprü de Muvaffakiyetle İkmal Edildi. Son Posta Newspaper. Page: 1.
58. Gönültaş, G. (1973 December 20–22) Interview: Türkiye'nin ilk kadın mühendisi Sabiha Rıfat'ın anıları. Milliyet Newspaper. Page: 4.
59. Nafia Vekaletinden (1933 August 4) Tender announcement. Vakit Newspaper. Page: 12.
60. BCA 19.02.1942 dated addendum to the contract, 30-10-0-0/Muamelat Genel Müdürlüğü-Ordu-Fatsa yolundaki Bolaman köprülerine ait ek sözleşme hakkında Danıştay'ın görüşü.
61. BCA 31.12.1930 dated. 30-18-1-2/Kararlar Daire Başkanlığı (1928-) Kilis-İslahiye Demiryolu arasında yapılacak Aferin köprüsünün malzeme ihtiyacı için kambiyo alınması.
62. Project Works (1936, July 1). Bayındırlık Works Magazine, Administrative Part Year: 3/2. Page: 118.

63. Progress of Works (1937, July 1) Bayındırlık Works Magazine. Technical Part. Year: 4/2, Page: 7.
64. Progress of Works (1937, July 1) Bayındırlık Works Magazine. Administrative Part. Year: 4/2. Page:106.
65. Cumhuriyetin 15 Yılında Türkiye Bayındırlığı (October 1938). Bayındırlık Works Magazine, Administrative Part Year.5, S.5. Page 337.
66. Haykır, Y. (2011) Atatürk Dönemi Kara ve Demiryolu Çalışmaları. Doctorate Thesis submitted to Fırat University. Page: 661.
    The existing road shown was soil stabilized road from Elazığ to its north crossing over Fırat River in as shown in 1931 dated map. The crossing over Murat river, which joins Fırat, was provided at Pertek location with ferry. The oldest bridge known in this vicinity is the 1934 dated timber bridge constructed by Military.
67. Elazık ve Tunceli Halkı Büyük Şefi Emsalsiz Tezahürlerle Karşıladılar (1937, November 18) Ulus. Page: 1.
    Other Newspapers reporting the visit:
    Atatürk Dün Elaziz den Tunceli ne Gittiler (18.11.1937, November 18) Akşam. Page: 1.
    Atatürk Elazızde (1937, November 18) Cumhuriyet: Page 1.
    Atatürk Şark Seyahati İntibalari (1937, November 22) Cumhuriyet. Page 1.
    Ataturk dün Tuncelini Şereflendirdiler (1937, November 18) Son Posta. Page: 1.
68. This 1937 dated bridge was a multispan timber bridge with trestle bents.
69. BCA State Republican Archives, 07.04.1938 dated 30-18-1-2/Kararlar Daire Başkanlığı (1928-) Şile-Ağva yolu üzerindeki Göksu köprüsünün ilave inşaatı olan betonarme işinin pazarlıkla eski müteahhide yaptırılması.
70. Köprüler (1939, November 1). Bayındırlık Works Magazine. Administration Part. Year: 6/5.
71. Akkaya (1989) Pages: 78–81.
72. His method works on the same principle as the device used today to measure "acoustic strain gauge". These devices measure the frequency changes of the wire whose length and natural frequency change under the load.
73. As heard KGM was planning to restore and repair the Akçay, Meşebükü and Miliç (BWS-18) bridges
74. Figure 4.27 Drawings of the reinforced concrete arch of the Pertek Bridge, by Prof. Mörsch, in 7.07.1937 Source: State National Archives-Republican Archives, Binder no: 2319. Found in Örmecioğlu, H. T. (2010) Page: 140.
75. Mühendis Feyzi Akkaya'yı Anma (8 January 2005) İnşaat Mühendisleri Odası İstanbul branch Narrated by anonymous engineer. Maya Basın Yayın. İstanbul.
76. **Davut Parker (Pistiryakof)** was a Russian immigrant. While working as an engineer in the service of Tsar Nicholas, he had to migrate to İstanbul due to the October revolution. After working as an engineer and subcontractor for a while, he started his own contracting business. His name Davut given to him by Atatürk, instead of David, and he deleted his surname with a court decision. His office was next to Akkaya's office in SAFERHA. Further information can be found at Chapter 06 of this book: Sırzı MTL-09 and Kozluk MTL-10 Bridges.
77. Huguenin Köşkü (2016, 26 September) Interview with Madam Eteri Pincas Parker. Retrieved from: http://www.gdd.org.tr/mutfakdetay.asp?id=217
78. The mansion the Parker family stayed was constructed by Edouard Huguenin, The General Director of the Anatolian Railway Company between 1908 and 1917. After war, family bought the mansion. This heritage listed mansion, close to Haydarpaşa railway station, shown to Atatürk when he was looking.
79. BCA State Republican Archives, BCA 030_10_00_00_155_92_2 Murat nehri üzerine yapılacak 145 m uzunluğundaki betonarme köprünün kalıplarının tamamlandığı.
80. Pertek Köprüsü Yakinda Açılıyor (1939, July 5) Akşam Newspaper. Page: 1.
81. Akansel, C. (2009) Revealing the values of a Republican Park. M.Sc. Thesis in Architecture. Middle East Technical University.
82. Özdil, N. C., Vejre, H., and Bilsel, F. C. (2020). Emergence and Evolution of the Urban Public Open Spaces of Ankara within the Urban Development History: 1923 to Present. 26–51. https://hdl.handle.net/11511/47660

83. **Jacques Nessim Aggiman:** Born on March 7, 1892 at Monastir, Turkey, he received his engineering education at McGill University, Montreal, where he graduated as a B.Sc. in 1917. He worked as a draughtsman from 1911 to 1915 with the St. Lawrence Bridge Company, at Montreal, on the Quebec bridge project. Then he lived in Turkey and worked as a contractor. In 1942, he moved to America.

Zelef, M. H. (2017). A Hidden Figure in the Construction of Embassies in Ankara: Jacques Aggiman. *Journal of Ankara Studies.* DOI: 10.5505/jas.2017.97769.

84. Akkaya (1989) Page: 100.

85. Nafia News (1937, September 13) Ulus Newspaper. Page: 4.

86. Akkaya (1989) Pages: 83 and 98.

87. Reşat, Y. (2017 November 21) Bu köprüyü sel vurmuştu! Cagdaş Web TV. Retrieved from: https://www.batmancagdas.com/yasam/bu-kopruyu-sel-vurmustu-h54322.html

88. Nafia (1939) Köprüler. (Administration Part) Year: 6/5.

89. The photographs for Alikaya and Suçatı bridges are kindly donated by Prof. Durmuş Öztürk.

90. The information about the current condition of the bridges are kindly provided by Yusuf Köleli from https://marasavucumda.com/. He also has supported the search for Maraş bridges and provided the photograph of Tekir bridge.

91. Akkaya (1989) Page: 83.

92. Nafia Tender Announcements (1936, August 30) Son Posta, (1937, March 12) Kurun and (1937, April 25) Akşam in the same order.

93. Akkaya (1989) Pages: 83 and 185.

94. Güney, H.N. (2020) *Ordu'nun Yol ve Köprü Hikayeleri.* Ordu Büyükşehir Belediyesi. Ordu. Page: 201.

95. ***Orhan Büyükalp*** was born in 1925 in Maraş. He graduated from ITU Faculty of Civil Engineering in 1950. He worked in Diyarbakır Ninth Regional Directorate of Highways until 1959. During his military service, he served as the founder of Antalya Highways 13th Region, where he was the directorate for 14 years. He worked in different Regional Divisions of Highway Directorate until his retirement in 1982.

96. Meslekte 40 Yıl (1990 December) *Türkiye Mühendislik Haberleri.* IMO. Ankara.

97. ***Muammer Kiraner*** one of the well-known contractors of those years had won the tender for the Urfa-Viranşehir-Kızıltepe-Nusaybin-Cizre road from the Highways.

98. No name. (1956) Birecik Köprüsü. Arkitekt Magazine No: 1956-01 (283) Pages: 19–20.

99. This document could not be found in archives.

100. Akis ("echo") was the first of the news-politics magazines in the Turkish press published 706 issues in total between May 15, 1954 and December 31, 1967. It was issued under the direction of a prominent journalist and writer, Metin Toker.

The detailed and through, especially in economic terms, text had been shortened and translated from Akis Magazine dated 7 April 1956 retrieved from the archive of İnönü Foundation at http://www.ismetinonu.org.tr/.

101. Birecik Köprüsü (1971, July 06) Cumhuriyet Newspaper. The photograph of the tomb also given. The story has many versions and each version makes sense and can be accepted a part of truth. As learned from social media the tomb was removed from this place. Page: 4.

102. Ağaoğlu, H. (2018) *Karayolcu: Bir Mühendisin Anıları.* Everest Yayınları. İstanbul.

103. Akkaya (1989) Page: 200.

104. https://kantaratlas.blogspot.com/2015/08/hasankeyf-bridge.html.

105. Tender Announcement (1960 March) IMO News Bulletin. Ankara.

106. Ağaoğlu, H. (2018).

Chapter 6

Metal Bridges

6.1 INTRODUCTION

Metal bridges in Turkey are extremely rare for the period of the study, the Early Republican era. The main factors contributing to this include that iron production was low in volume and under government control in the empire and thus it was expensive, discouraging the use of metal and its development.

Metal has been applied to all bridge types and often led to further improvements. For example, concrete truss bridges had been experimented with for a while, but after a few attempts, they were abandoned, as they could not compete with the potential and variety offered by steel truss bridges. Metal bridges are a broad category and include almost every structural type: beam, arch, truss, pontoon, suspension and finally a movable bridge type. Among them, the truss is the structural type used most commonly and will be referred to here as the main type for metal bridges.

In a basic truss bridge, the main members, which also define the envelope of the structure, are the lower and cupper chords. They are connected with vertical and diagonal members. At the deck level of the truss, where members constitute the deck for traffic to be carried, the longitudinal beams, called stringers are used to carry or transmit the load to chords via transverse beams, also called secondary beams. As the truss does not have any stabilization in the transverse direction other than the connecting deck, additional transverse members are needed at the upper level for maintaining stability against lateral loads. The elements used at the entrances of the truss are also defined as "portal" and "end" in reference to their locations. The truss structure is composed of repeating arrangements between vertical posts, called panel or bay (Figure 6.1).

6.2 HISTORY OF METAL BRIDGES IN TURKEY

The use of metal goes back to ancient civilizations; however, its use in bridges has a specific timeline, like the first cast-iron bridge of Coalbrookdale in 1779. This bridge was a transitional structure; it was designed using the arch principle, familiar from masonry bridges, which works in compression and was also detailed with timber style joints. Then, between 1800 and 1900, there was a transition period from cast- and wrought-iron bridges to mild steel which was first used in 1874 for the Eads bridge over the Mississippi River by James Eads.

DOI: 10.1201/9781003175278-6

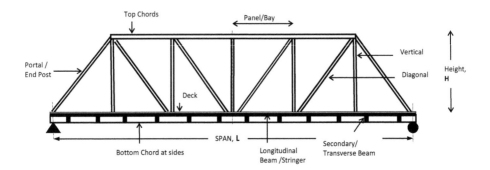

Figure 6.1 **Truss bridge terminology (author).**

The joint type is also a significant feature and indicates the date and characteristics of its period for the bridges. The early bridge connections were made with rivets, which demanded experienced labour. In the USA, pin joints were also used. Later bridges used bolts and welds with the advancement of technology.

Metal, lead and iron were used in Turkey in the construction industry since Ottoman times. Its structural use can be observed in lead clamps for masonry connections, ties for the arches of domes, brackets for cantilevers, the lead coverings of domes and the nail for timber construction.

Metal production during the Empire was mainly for military purposes such as producing cannonballs, sword parts and bayonets, and the first cannon production yard was established in Tophane, with the name derived from armour, built in the 15th century. The first iron factory was the 1842 Zeytinburnu foundry with iron tubes, steel rails and other metal products manufactured; however bridge parts, which needed special moulds were not available.[1]

In the early republic period, small spans of about 20 m were available in the local market since 1926 from Eskişehir Factory.[2] This factory was initially established as a workshop for steam engine repair and maintenance in 1894 by Germans for the Bagdad Railway. The workshop expanded and improved with additional facilities, including the manufacture of bridge beams and other railway system parts and owned by the Republic since 1920.[3] This factory also produced the first automobile and first locomotive, and they were named "Devrim" and "Karakurt", meaning "revolution" and "Black Wolf", respectively. Both were locally produced.

MKEK,[4] the first steel factory of the republic, started production in 1932 in the Kırıkkale province. When rails were first produced in 1932 and submitted to the authority, they were not blindly approved. As Sanbaşoğlu[5] narrates: "When we first delivered the rails, Turkish State Railways (TCDD) was sceptical. İsmail Fuat Bey, who was the Head of the Construction Technical Committee, took a few samples and sent them to Switzerland for testing". This is good evidence to demonstrate the independent actions of engineers and authorities.

Then, the "Karabük Iron and Steel factory" started operations in Karabük in 1939. This was the first industrial complex of the Republic and provided steel for nearly the whole local market. The factory was coal-powered, fed by a mine located in the vicinity and connected via a railway line. This factory was opened in 1937 by Prime Minister İsmet İnönü, and as a tradition, the ovens were named with female names like Fatma and Zeynep to imply productivity.

The first steel bridges in Turkey were imported for Railway lines or roads to service the railway lines which were constructed through foreign companies. Common construction materials at the time were stone, wood and metal. Stone and timber were locally available; and were often free through the concessions, which gave access to resources on both sides of the railway. Steel was produced in other countries and imported free of tax and favoured as there was no need for formwork and construction was quicker, practical and independent of local conditions. Structural steel was also the main material for railway bridges around the world, due to its greater strength, ductility and fatigue endurance.

Many road projects were also contracted to foreign companies through finance provided by them as well. These road projects were mainly intended to feed the railway lines and the harbours for the convenient import and export of raw materials and goods in Anatolia.

The earliest metal road bridges built within the current borders of Turkey were the bridges in Kars, at the eastern border of Anatolia which was under the rule of Russia during 1877–1917. During this period, Russia constructed railway lines and roads for military and economic reasons. KGM inventory gives 24 of these but a detailed list was not available.

Çamçavuş and Kağızman bridges are mentioned in related publications. These bridges have distinctive features such as a curved upper chord, timber decking and double bay diagonals. The main chords are T-section made with riveted plates, the diagonals were L-sections and the verticals are detailed with outward cantilevers. Some bridges had metal substructures with circular column piers and truss piers with an enlarged base. Çamçavuş bridge had a 60 m span with a 7.5 m height truss. The bridge was originally located over Susuz creek; however, since it would have been submerged in the lake of a nearby dam, in 2014 the bridge was relocated 15 km away to serve as a pedestrian bridge for Kafkas University. In its new location, the bridge is crossing the Kars River. The structure, weighing approximately 100 tons, was transported in one piece by truck to preserve its original state (Figure 6.2).[6]

Metal bridges were also used in specific projects in the capitol of the Empire, İstanbul. An interesting collection of iron bridges are used for landscape design in Yıldız Palace Garden for crossing over artificial channels.

Yıldız Palace is in the Beşiktaş district of İstanbul. There is an artificial canal and lake in the Palace gardens. As the stream looped around the garden, there emerged an island on the south side of the garden. The Island had some animal shelters and bridges that made it possible to control the animal's movements. This interesting movable bridge was made of steel and operated with a rod to open and close the span. It is a bascule type bridge as seen from its photograph (Figure 6.3).[7]

Golden Horn crossings in İstanbul are another important subsection within the history of Turkish metal bridges. The first crossings were of timber, and after 1875, metal bridges were used for Galata and Unkapanı locations. Bridges always had movable parts in the middle with swing or bascule type openings. 1940 dated Gazi bridge, later named Atatürk bridge, is included in this chapter with detailed descriptions. For

Figure 6.2 Kağızman bridge with its double bay diagonals (author's collection).

Figure 6.3 Bridge-I bascule type movable bridge connected the island to mainland (İstanbul University Library - rare works collection).

others, a thorough list is provided in Chapter 2 of this book to summarize their technical features.

Nearly all of the Empire and early Republican metal bridges had their superstructures brought from France, Germany, Czechoslovakia, Belgium and the UK. The foreign companies benefited from tax exemptions for all the imports relating to machinery. Since the Ottoman times; vehicles, implements and materials required for the bridge building were imported with the specific approval of the government.[8] The same procedure was maintained during the early Republic; however, there seems to be a formal approval for each case with the new authority. The equipment was then usually returned back to their origin countries after being used for the assembly.

A variety of companies from France and Belgium supplied the road bridge design drawings and the material. These companies probably were local firms subcontracted by investing companies. Below are some of the titles in the drawings of the projects found in state archives (Figure 6.4).[9]

There is no definite date that marks the start of the local production of steel bridges. For example, the Galata bridge built in 1992, which had the longest bascule span of 80 m when constructed, had its steel imported from abroad. Whether the need for import by then was due to the inability of the local market to fulfil the needs or the requirement of the project financing model for this particular structure is not known.

Steel bridges were not common, especially for road bridges. The main reason was the higher initial cost and the ongoing maintenance requirement during the life of an asset to avoid the corrosion of steel, compared to its alternative concrete. Therefore, steel is considered expensive in comparison to other materials and is only used whenever there is a solid justification. For example; eliminating the piers with longer spans, reducing the load on foundations or decreasing the structural height of the superstructure would make the use feasible. Steel bridges were also used in challenging situations like topographical conditions impeding the construction activities or the remote areas

Figure 6.4 Drawing title blocks from various metal bridge projects in Ottoman archives. Numbered 1–5 from left to right (BOA-PLK-1531, 4164, 1377, 5577, 2141).

with limited access. This resulted in some road bridges with remarkable technical features like Borçka Bridge with an impressive Langer type arch and Kemah bridge with suspension cable and stiffening truss deck and the Bartın bridge with bascule span.

Another interesting example which is currently still in use is the Kirazdere bridge in İzmit, constructed from the leftover steel of British gun emplacements from World War I.

6.3   TECHNICAL FEATURES OF METAL BRIDGES IN TURKEY

Material Strength: A standard for bridges published in 1938 did not include steel as a material of construction and no reference was available for its use. However, this standard included reinforcement for concrete specified with a 3700 kg tensile strength and referred to German standards for other parameters. The previous 1933 Standard had not even mentioned reinforcement.

A 1940 dated book for Railways by Meissner[10] described the steel used at that time:

> Until 1922, steel with 37 to 44 kg/mm² strength was used for nearly all the construction works. In exceptional circumstances, special steels were used with a content of 2.5 to 3.5% of chrome or nickel ... Since 1925, a steel with 48 to 58 kg/mm² strength was adopted and called St48. Later, stronger steels with nickel and silicon content were made available.

Peynircioğlu 1940[11] also provided consistent information and recommended to use high strength steel due to "its total lightweight for structure dead load, and implying lower transportation and assembly cost, and reduced dimensions and connections than its predecessors".

Superstructure Types: Metal bridge superstructures were mostly plate girders or trusses. Plate girders were used for spans up to 30 m, and beyond this range, truss types were employed. In addition to these two common types, there were, for example, arch bridges of Borçka and Ceyhan, a suspension bridge at Kemah, deck arch bridge for Sırzı, Bartın bridge with a bascule span and the pontoon Gazi bridge. Ottoman bridges, except the Yıldız Landscape bridges, had all been girder and truss types (Figure 6.5).

The following information and observations are taken from the archived road network projects of the Ottomans and early Republic, which form the basis of engineering in metal bridges in Turkey.

The projects mentioned also provide an insight to bridge projects in Ottoman Nafia.[12] Projects were submitted in French as it was the language used for foreign affairs. Their date ranges are between 1901 and 1916.[13]

Deck Types: Typical bridge widths used in road bridge projects were 5–8 m, with or without sidewalks. Usually, a narrow path for pedestrians on both sides of the section is provided as cantilever extensions. The allocation of the width for the traffic lanes is often not clearly shown. Indeed, the road section should have been determined by the authority, as the asset owner. However, the details in terms of traffic lane configuration varied from bridge to bridge.[14]

CHEMIN DE FER DE MOUDANIA A BROUSSE (ASIE MINEURE)

اوغزرشتی فیلعززوره نیلوفرازریدیانلعی تریفسدورسه سانتمطولله ولوری

PONT DE 82ᵐ.74 SUR LE NILUFER AU KILOMÈTRE 35

Figure 6.5 **One of the girder bridges used in the Bursa-Mudanya Railway line (İstanbul University Library - rare works collection).**

The deck for steel bridges is the main challenge as the steel members do not provide a continuous area for wheeled loads travelling along the road bridge crossing. The deck configuration, determining how to overlay the comparatively slimmer transverse beams, is an important feature as it affects the cost and the construction practice. Moreover, the live load should be distributed safely and uniformly over the beams.

Generally, metal decking is made by zores,[15] which are sections laid on the secondary beams and fixed by bolts. Concrete is poured over the zores sections, followed by a sand layer and a macadam[16] pavement finish used for the road surface (Figure 6.6).

Concrete decking is normally made by a reinforced concrete layer placed on top of transverse beams, which have the upper flange embedded in the concrete, followed by a road finish.[17] In this case, the bridge section does not show any connectivity between steel beams and deck so the composite action is not attained.

In the case of bridge decks made of timber, thick lumbers are laid on transverse L-sections. A small gap is usually kept between the lumber and transverse beam to protect timber by allowing the water to drain away.[18]

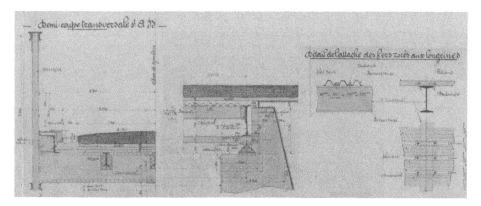

Figure 6.6 Metal bridge over Menderes river detailed with metal decking dated 1914 (BOA-PLK1531).

Standard configurations were prepared for different deck widths as 6.5, 7 and 8 m. Superstructure main beams were I beams and between them, so-called "buckle plates" are utilized. Buckle plates were made from wrought iron and curved upwards (hogging); in some cases, plates that curved downward (sagging) are also used (Figure 6.7).[19]

In various design drawings, concrete or metal decking was shown with a road surface of mostly macadam with occasional asphalt and timber applications. Drawings did not give specific project names, and rarely named the rivers crossed; instead, these drawings were titled as "typical" details.

The beam sections were mainly solid rolled I-sections up to a maximum of 304 mm height; beyond this height, a compound section is used. Formed sections were made of plates joined together to make I beam shapes with L and U profiles at corner connections. For example, the built-up steel members for Gediz and Yahşihan bridges were made with plates connected with L-sections. Yatağan, Manavgat, Vezirhan and Genç bridges had significant improvements in terms of member sections as fabricated U-profiles were used together with L-sections in combination to form the built-up sections of the laced and battened system (Figure 6.8).

Assembly: For the imported steel superstructures, foreign manufacturers were also responsible for the assembly on site. There is a document[20] for the safety and assembly procedures shown with notes. This instruction sheet prepared in French does not have a date. It is not specific to bridges and describes a safety belt and securing the belt to a pole, erection procedures and preparation of steel structures and poles for painting (Figure 6.9).

It is evident from the documents in the archives that these firms sent their own engineers and technicians for assembling the bridge parts. From KGM archives,[21]

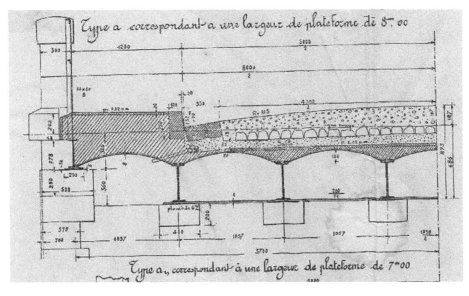

Figure 6.7 Sections with a decking form made of buckle plates (BOA-PLK-1377).

Figure 6.8 Built-up sections details from left: Yahşihan, right: Manavgat bridges (author).

the name of the professionals who worked as "mounter" engineer for the installation works is documented. The term mounter, sometimes monitor, was used for the technicians working in the assembly works.[22]

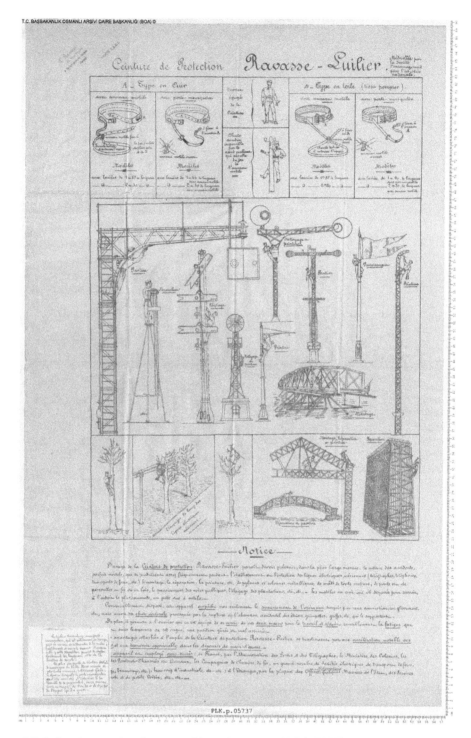

Figure 6.9 Safety instruction for assembly and painting (BOA-5737).

Atayman,[23] an Ottoman Nafia engineer in Kosovo between 1907 and 1910, narrates the assembly of a bridge and provides descriptions of the work. This 40 m long bridge was a two-span truss iron bridge with spans of 20 m. It was located in the Pristine district in Serbia (later Yugoslavia) and crossed over the Lab River.

> Before winter, the masonry substructure of the bridge was completed by the road contractor. Later, during the most severe time of winter, the bridge parts, which were ordered from Germany, and three monitors for the assembly arrived at site. After the iron parts were moved from the Pristine [rail] station to the bridge location, we were sent to work with three German monitors to begin the assembly of the bridge…
>
> The point that I want to emphasize here is the fact that the monitors continued their work with diligence and endurance under the severe cold that prevailed there during December and January. The installation of the bridge with two spans of 20 m was completed in two months with three monitors and 5 to 6 Albanian workers, who worked always as a team, without a break, even on Sundays.

Joints: From available accounts, riveting was the joining means for nearly all the bridges in the Ottoman and Republican Nafia even including the 1952 dated Genç bridge.

In the 1930s, riveting was beginning to be replaced by welding and bolts started to be used in bridges in the 1940s, internationally.

Riveting was labour-intensive work and needed at least 2–3 workers. Rivets needed to be preheated until they were, literally, red hot, so that they could be hammered and moulded through the predrilled holes. The worker needed to catch the hot rivet and insert it at one side of the hole, and another worker to hammer it down to a round head. Atayman describes the riveting for a Lab River bridge in Pristine: "I was watching the two people concurrently hitting the rivets of the iron parts, which were placed in their locations after these became crimson red heated with mobile bellows".

Substructure: Bridge substructures were predominantly built with masonry. The use of concrete for the substructure is not observed until concrete became economically viable. This may be assumed to have happened after 1910,[24] which was the year when the cement factory started operations in Turkey. Substructures for metal bridges were mostly constructed by the road contractors.

At the same time, new applications were being introduced in bridge building via the designs by foreign companies. The use of screw piles is one of the innovations, which was invented in 1836 and regarded as one of the significant advances in engineering.[25] Screw piles were placed in a triangular arrangement in plan and the individual posts are extensions of these projecting above the ground level, forming a "Trussed pier".

In a book dated 1873,[26] screw piles are described as a solution for foundations on soft ground. 'These piles have screws of different forms according to the soil they are to be used in. The point being little or nothing, and the thread of the screw very broad, for loose soils; the point becoming sharper and the thread of the screw narrower as the soil becomes harder.'

Figure 6.10 Gediz bridge photograph (İBB Atatürk library).

Trussed pier and screw piles found their way to some bridges in Turkey. One of the known examples is the 1903 dated Gediz bridge with six spans of simply supported trusses with a parabolic upper chord.[27]

The Gediz River is Turkey's second-longest river with its 401 km length in Anatolia flowing into the Aegean Sea. This region was connecting the Smyrna hinterland with a harbour through the first railway constructed in 1856 between İzmir and Kasaba (Figure 6.10).

Atayman[28] endorsed the use of screw piles for bridge piers in 1908, describing them as: "... preferred option for an iron bridge resting on screw piles for urgent projects to finish in quick time and less money". He also made a comparison to the piers of the Yahşihan bridge

> It would take at least 2 years for the masonry bridge to be constructed safely and successfully. However, it was possible to build an iron bridge of this size on screw iron piles for a short period like four months...

Atayman does not provide information on how the pile is inserted into the soil.

A set of drawings provides details for a leg frame with calculations.[29] This 1903 dated project for the construction of the Gediz bridge shows that the pier is made of piles driven into the river to form a truss system. In drawings, the hollow circular piles are shown with 12 cm diameter with screw blades 80 cm in diameter (Figure 6.11).

The benefits of the truss piers and screw pile system are the ease and practicality of its construction and especially not requiring special machinery. At the same time the spread of the piers to the enlarged area in the plan, less disturbance of the river flow and catch less debris, would also make the design a favourable option. This design is repeated in Çorlu and Gediz bridges as well and also later in the Güzelhisar (BRD-16) bridge, and concrete piers were designed using the same design feature (Figure 6.12).

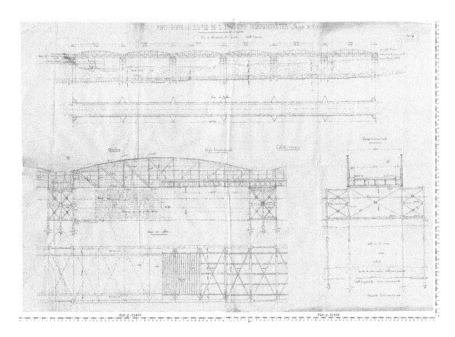

Figure 6.11 1903 dated Gediz bridge drawings (BOA-PLK-1400).

A 1943 dated book in Turkish by Peynircioğlu[30] describes piers made with screw piles as "unsuitable for permanent bridges", referring to the truss pier. This pier system is prone to collapse if just one truss member is damaged or failing. Moreover, the whole system is prone to corrosion requiring ongoing maintenance.

The use of screw piles continued only until 1920[31] and was replaced with concrete piling especially to overcome the durability issues.

Quality Control: Örmecioğlu[32] discusses the introduction of quality controls after the difficulties that occurred during the construction of the Manavgat bridge: Because of the problems that occurred during assembly, the need for controlling the production process was realized. For future projects, the Department of State Railroads would recommend Robert W Hunt Co. from England as a controlling firm, for the quality control and the testing of the production of steel elements of later bridges made of steel, such as the Borçka and Ceyhan bridges on behalf of the Ministry of Public Works.

A 1943 dated document[33] from the Republican archive mentions Robert Hunt Company as an agent from 1928 for the provision of services relating to quality control and procurements from England and many other European countries for the State Railways.

6.4   METAL BRIDGES IN TURKEY

As the history of metal bridges commenced in the time of the late Empire, Ottoman public works bridges are also mentioned in this chapter.

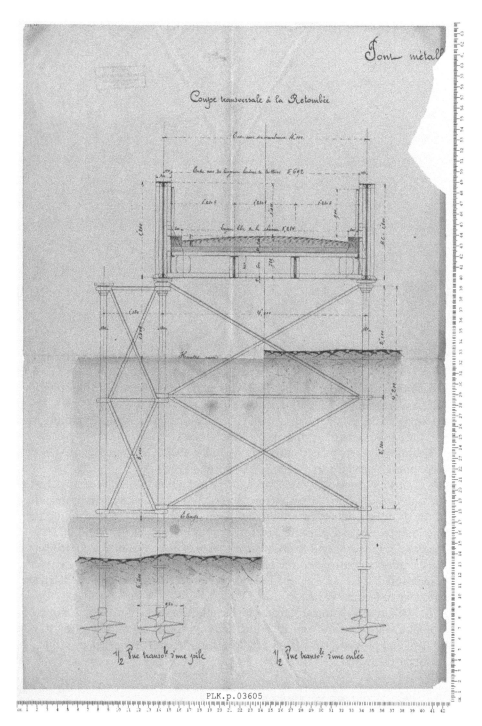

Figure 6.12 1903 dated Gediz bridge and 1906 dated Yahşihan bridge pier details with screw piles (BOA-PLK-3605).

Metal bridges built during the Ottoman Nafia can be grouped under three categories: Railway bridges, İstanbul Bridges and Road bridges. Railway bridges are not within the scope of this book. İstanbul bridges were mainly for Golden Horn and Bosporus strait crossings. These bridges are discussed in Chapters 2 and 3 of this book. A subsection of İstanbul bridges is the landscape bridges in Yıldız Palace gardens dated 1894–1895 as briefly mentioned in this chapter.

In terms of road bridges, which were mainly constructed to complement the railway network, the Ottoman government engaged a French Company, Regie,[34] for the construction and repair of these road networks, through a special contract in 1909.[35]

Therefore, Ottoman archives have a significant number of projects prepared in French during the period 1901–1916. In addition, many bridge projects undertaken in previous years[36] were also available as foreign engineers were engaged in most of the engineering works in the Ottoman Empire.[37]

Some metal road bridges from the Ottoman period were still in service during the early republic era. Two of them are the Yahşihan bridge built in 1906 and the Yatağan bridge built in 1915.

Yahşihan Bridge is in Kırıkkale, crossing over Kızılırmak (Halys) River. It has four truss spans in total, three of them are through trusses each 26.95 m in length and the fourth span is a simple beam. The total length of the bridge is 99 m with a deck width of 4 m. The construction of this bridge has been detailed by Nafia engineer Atayman,[38] who gives some insight into the conditions at the time.

Governor Cevat Bey queried Atayman about the possible means of crossing Kızılırmak (Halys) river for travelling from the West to Ankara in 1905. His reply was the existing raft run by 4–5 people in Yahşihan with a pre-determined cost per person, carriage or animal. The raft was pulled by ropes or pushed with poles. However, for the local farmers living in the area, it was too costly to pay for regular use.

The Governor then revealed his decision to construct a bridge and gave orders to collect donations from nearby villages. At the same time, Atayman was employed to investigate the bridge site, determine local resources of stone, sand and lime for construction and prepare the project and bill of payments for the bridge. Atayman completed the work and determined that 20.000 Gold Turkish Lira (GTL) was needed for a 100 m long stone bridge. As the bridge was considered an urgent need, the nearby villages were requested to donate funding for the construction of the bridge. Donations were collected for the next 3 months locally. In total, 3.000 GTL was accumulated.

As the money was not enough and the bridge was needed urgently and was not possible to collect the remaining amount, the conditions were reported to Nafia. Nafia added another 1.000 GTL and ordered an iron bridge from a Belgian company at a cost of 4.000 GTL. The bridge was fabricated and assembled by the same company within 4 months.

The abutments were constructed by local road contractors not included in this total.

Atayman provides further information on his duty for the transportation of iron parts to the bridge site. An adequate road for transportation was not available and it needed to cross over mountainous terrain. He was assigned the job of preparing a road as soon as possible as they had received a telegraph stating that the iron was ready to

Figure 6.13 Yahşihan bridge (author).

be shipped. He only had one kondoktör[39] and clerk with local villagers as workers. They had tents to stay in and only simple tools like pickaxes and spades and no budget to construct even a small culvert. The road was opened in about 20 days. The bridge parts were then delivered to the site and assembled by monitors [40] in 4 months and handed over to the province. The bridge opened to vehicles, livestock and military traffic in 1906 (Figure 6.13).

Ottoman archives have many drawings and design sheets dated between 1903 and 1907 for this particular bridge.[41]

One of the drawings shows the general layout of the bridge site giving the profile with water and ground levels. This drawing is in French and signed by the local authority as "measured, drawn and submitted by Ankara Province Head Engineer, on 5 September 1903, Y. Halim".[42]

After the war, many metal bridges were repaired by the Republic provinces. One of them was the 1915 Yatağan Bridge, connecting Muğla to the Aegean coast.

Yatağan bridge is a single span truss in Yatağan district of Muğla province, crossing over Dipsiz (Yatağan) Creek. The span is 40 m with a 5 m width as shown in drawings found in the archives.[43]

The photograph of the bridge is found in the Province Works Album[44] and information on the photograph provides the construction date of the bridge as 1915.[45] The name of the town of Ahiköy was changed to Yatağan in 1945.[46]

In 2005, half of the bridge collapsed, when a truck loaded with marble hit the upper bracing of the truss. The bridge had to be quickly repaired with temporary support. The remaining 16 m of the bridge was supported on a temporary pier and the collapsed section replaced with a narrow span serving only pedestrians. The bridge was completely restored in 2013 (Figure 6.14).

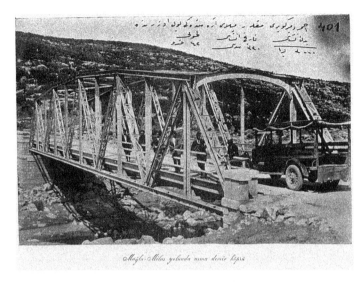

Figure 6.14 Yatağan bridge photo from Province album (İBB Atatürk Library).

6.5   EARLY REPUBLICAN METAL BRIDGES IN TURKEY

The first steel bridge of the Republic era was the Dalama bridge dated 1931. There were in total 17 steel road bridges known to be constructed from 1923 to 1966, and 14 of them are described individually in this chapter (Figure 6.15).

Metal was not a common material for use in road bridges and was only chosen where necessary. Therefore, metal road bridges built during the Ottoman Nafia years were 11 in total. Eight of them[47] still survive today.

The most common structural type used was truss; however, other types such as Langer arch, deck arch, pontoon, girder and suspension structures were also observed mostly as the first and even only representative of the type in Turkey.

Since the total number of these bridges is so low, each of them can be regarded as significant in terms of representing the structures of its type.

The list of these bridges is given in Table 6.1 with some information such as name, completion date, location, obstacle crossed (mainly river), superstructure type, number of spans, length of main span, current condition, supplier for the superstructure, contractor and coordinates (Table 6.1).

The designs for the Dalama and Vezirhan bridge superstructures, dated 1931 and 1932 respectively, were available and adopted from earlier projects of the years of the Ottoman Nafia. There is no record for the Dalama bridge for its current condition.

Therefore, the Manavgat bridge is really the first bridge designed and constructed by the early Republic. This bridge is still standing next to the new road bridge and now serves for pedestrian use. It is in a fair condition and is considered significant as its main structural elements and details are all original.

Vezirhan Bridge is also in fair condition although abandoned, and it can be spotted from the main road network.

Borçka Bridge is an outstanding example for its unique type and the structure is in good condition adjacent to the main road, still in use for pedestrians.

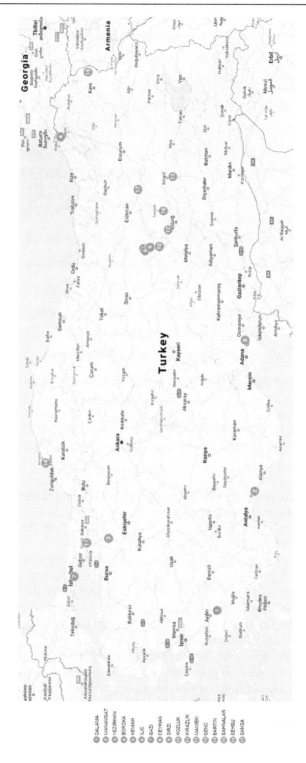

Figure 6.15 Republican metal bridges locations shown on Google Map (author).

Table 6.1 List of Early Republican Metal Road Bridges in Turkey (author)

MTL No	Name	Comp. Date	Location	River	Structure Type	No of Span	Main Span Length (m)	Status	Substructure/Contractor	Coordinates
1	Dalamaçay	1931	Aydin	Dalama/Madra	Truss	3	27.6	Incomplete information	French Company/SA-FER	37.67651, 27.99101
2	Manavgat	1932	Antalya	Manavgat	Truss	3	60.95	Standing	Flander AG Ali Nuri Bey/Izmirli Hayri Bey	36.78715, 31.44781
3	Vezirhan	1932	Bilecik	Sakarya	Truss	2	45	Standing	TCDD/Vahit Bey	40.23386, 30.06581
4	Borçka	1935	Artvin	Borçka	Langer arch	1	113	Standing	Anciens Etablissements Skoda Karl Anders and Dr. Cudi Nusrat Bey	41.36188, 41.67553
5	Kemah	1937	Erzincan	Firat	Suspension	1	52.8	Half Submerged	Raşit Börekçi	39.43503, 38.45174
6	Iliç	1937	Erzincan	Firat	Truss	3	36	Collapsed	Raşit Börekçi	39.40161, 38.38421
7	Gazi Ataturk Unkapani	1939	Istanbul	Haliç	Pontoon	24	19	In use	Man/Safer	41.0421, 28.96508
	Sahnalar	1940	Kars	Arpaçay	Truss	1	38	Incomplete information	Muhtar Arbatli Ve Ortak	40.73724, 43.55127
	Şehsu	1941	Tunceli	Munzur	-	1	70	Incomplete information	Seibert GmbH Fasih Saylan/ Salih Sabri Taslicali	38.979, 39.53581
8	Ceyhan	1941	Adana	Ceyhan	Arch	4	80	In use	Man A.G. Hugo Hermann/SA - HA and Ziya Baglar	37.03208, 35.81118
9	Sirzi	1942	Erzincan	Firat	Deck arch	1	40	Incomplete information	Witkovitzer Bergbau Alfred Schwarz/Davut Parker	39.28529, 38.4886
10	Kozluk	1943	Erzincan	Kozluk	Truss	-	-	Incomplete information	Witkovitzer Bergbau Alfred Schwarz/Davut Parker	39.07144, 38.5107
11	Kirazlik	1945	Izmit	Kirazdere	Beam	4	12	In use	Gun Emplacements From England	40.73544, 29.94165
12	Mameki	1946	Tunceli	Mameki	-	1	56	Incomplete information	Wagner Buro/Sabri Taşlioglu	38.85713, 38.98849
	Sansa	1949	Erzincan	Karasu	Truss	3	-	Incomplete information	Süleyman Betin	39.57602, 40.15064
13	Genç	1952	Bingol	Murat	Truss	3	55	Standing	TCDD	38.75124, 40.53008
14	Bartin	1966	Bartin	Bartin	Bascule	3	15	In use	STFA	41.67344, 32.2496

Kemah suspension bridge is exceptional for both its type and history; however, it will soon be flooded by the projected dam lake.

Ilıç bridge is known to have collapsed.

Gazi Bridge, still in service, is 82 years old serving vehicular traffic. This structure consists of metal pontoons.

Ceyhan bridge is another rare application of an arch bridge and is still carrying vehicular traffic.

Kirazlık bridge, made from gun emplacements from England, has scientific, historic and social value.

Genç bridge is a truss bridge and is now abandoned from the road network. One of the abutment embankments collapsed; therefore, it is accessible from one end only. This bridge has vital importance to the history of the area.

Bartın, the movable bridge, is still operating. This bridge can be regarded as an open museum both for its technological and historical characteristics.

For metal bridges, the defining features include main and secondary structural members, the connections used and also the supports which reveal the technology of its time. Moreover, in Turkey, any metal bridge is also representative of an exceptional case which needed the steel even though it was more expensive. Therefore, all the bridges represented have heritage potential, not only as engineering structures but also for their representation of a certain period in history.

In addition, in many cases, the railings may also be categorized as character-defining features of these bridges. For example, Manavgat bridge still bears its original railings.

MTL-01: Dalama bridge

Dalama Bridge is on the Aydın to Muğla road. This 1931 dated bridge gets its name from the river it is crossing which is also named Madra in some documents and maps. The bridge was designed by a "French Company"[48] with the bridge superstructure imported to Turkey. However, the project was put on hold due to war. After WWI, the bridge was constructed as per its original design.[49] A map[50] found in the archives shows the Aydın-Muğla road starting km 0 in Aydın, crossing the Menderes (Meander) with a 120 m span bridge at km 9 and then continuing with a subsequent crossing over the Dalama river at km 33.[51] The Menderes bridge drawings were found in the archive;[52] however, Dalama bridge drawings could not be found (Figure 6.16).

The bridge substructure was built of masonry abutments and piers. This bridge had three spans each being 27.6 m long. The width of the bridge was 5 m with additional 0.85 m-wide sidewalks on both sides. The ground was of sand and gravel and was not very stable. Therefore, 7 m long concrete piles were driven under the abutments and piers with 31 and 26 piles respectively. The river was very prone to flooding; therefore, the foundations were protected with coverings of stone known as riprap (Figures 6.17 and 6.18).

The decking was concrete on zores steel profiles. Then waterproofing and protection layer was applied and the final pavement was 15 cm made of bituminous macadam. The bridge opened to traffic on 10 August 1931.

Dalama bridge was tendered together with Kalabaka and Menderes bridges. On 26th November 1929, the contract was signed by Engineers Sadık and Ferruh.

The bridge was not found standing in its indicated location (Figure 6.19).

Figure 6.16 **Dalama bridge from Nafia (1933).**

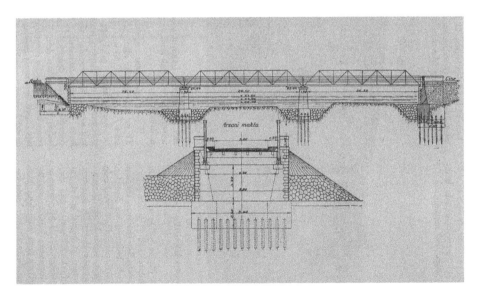

Figure 6.17 **Dalama bridge sketch from Nafia (1933).**

Three bridge project drawings were found in the Ottoman archives for the Aydın-Muğla road designated BOA 1531, BOA 3622 and BOA 3672.[53] For BOA 3622, only a drawing for substructure details was found. The BOA 1531 and BOA 3672 drawings show truss details, and although both are for the same length of 21.20 m between supports, the first truss is drawn with six bays and the second with eight, similar to Dalama bridge.

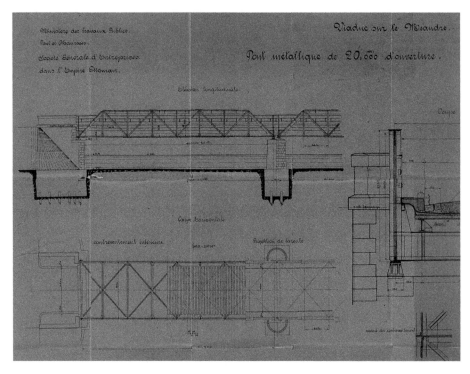

Figure 6.18 Detail drawings for a bridge similar to Dalama bridge from Ottoman archive (BOA-PLK-3672).

Figure 6.19 Construction of the bridge (KGM Calendar 1974).

MTL-02: Manavgat bridge

This bridge is on the road between Antalya and Alanya and is located in the town named Manavgat, which is also the name of the river. This is a coastal area and the bridge location is about 4 km away from the coastline. The depth of the river can reach up to 5 m at a high tide. Since the pier construction in such deep water would be considerably expensive, the pier was avoided with a main span made of steel reaching 60 m (Figure 6.20).

As Dalama Bridge was already commenced before the war, Manavgat Bridge can be considered as the first significant metal bridge of the Republic.

The difficulty in pier construction leading to a long span length and the advantage of being close to the road network facilitating the transport of the construction material to the site also favoured the preference of a steel bridge for this location (Figure 6.21).

After constructing the abutments and sinking the caisson of the pier on the Antalya (east) side, the pier excavation was started on this side. However, it was discovered that the rock bed presented fissures of concern meaning that it would not safely support the pier loads. Therefore, the pier was shifted toward the Alanya (west) side by 1.5 m. This resulted in the Antalya pier being shifted by the same amount. Then the caisson, which had already been sunk to 4 m deep was made redundant and the new pier was constructed with timber sheet piling. The ground was excavated to 3.5 m with subsequent timber piles being driven to support the foundation.

The design also includes outer concrete spans at both ends of the main span. The spans are 9.5 + 60.95 + 12.5 m starting from Alanya and going to Antalya (Figure 6.22).

A Decree[54] was found in archives for this occasion, formalizing the additional work required for the two pier foundations corresponding to an additional 20% of the project value. The justification of this revision was explained as "the lack of sufficient borehole or investigation". Two options were suggested for this situation. One was to cancel the existing contract and tender the remaining work again for others and the second was to tender the additional work separately. The second option was selected. The letter is signed by "Gazi M. Kemal" on 19 August 1931.

Although it would have been more practical to revise the superstructure to match the changes resulting from the unexpected ground conditions, it is inferred that the superstructure might have been already manufactured or on its way, therefore the revision was focused on the arrangement of the foundations.

The main span deck is made of concrete poured over zores sections.

The masonry substructure was awarded to the contractor on 09 October 1930 and the works were completed on 08 January 1932, taking approximately 15 months, presumably delayed as a result of the problems encountered with the foundations. The installation of the steel superstructure started on 11 December 1931 and the bridge was opened to traffic 4 months later on 03 April 1932, making a total duration of 19 months (Figure 6.23).

Örmecioğlu[55] reports issues related to the rivet connections that occurred for Manavgat bridge in a KGM Telegraph report due to misfabricated rivet holes on steel elements of Manavgat Bridge. This notice was submitted from the chief engineer Nihat to the Department of Roads and Bridges on 25 January 1932.

Currently, this bridge appears in good condition and is open for pedestrian crossing. The original railings and posts are still in place and the original abutment details

Figure 6.20 **Manavgat bridge section (author).**

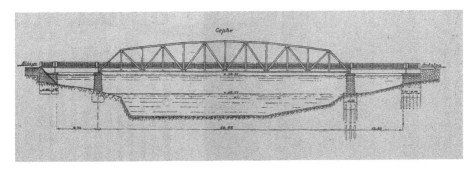

Figure 6.21 Manavgat bridge from Nafia (1933).

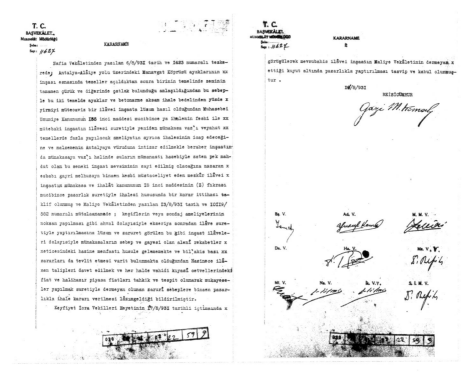

Figure 6.22 Decree signed by Gazi M. Kemal on date 19.08.1931 (BCA-22-59-09).

remain on the Antalya side. In regards to the opposite, Alanya side, it is occupied by street vendors, etc. and some parts appeared to be missing during the author's visit.

In 1999, the road was shifted downstream and another remarkable steel bridge was built. This new bridge was made of an orthotropic box section and constructed utilizing a cantilever method. It was the first of its kind in Turkey. Bridge had an 80 m span again avoiding piers in the river.

Figure 6.23 Manavgat bridge (author).

MTL-03: Vezirhan bridge

This bridge is located in Bilecik province between İstanbul and Ankara and connects the town centre of Gölpazarı to the Railway station in Vezirhan. It is crossing over the Sakarya river with two spans, each of 45 m with a masonry substructure.

This bridge was originally the old railway bridge crossing the same river downstream to the north. The bridge had been used for the İzmit-Ankara railway line which was constructed during 1888–1893 by a German company. It was named Lefke bridge and it is known that it was damaged during the conflict, subsequently being repaired by an Engineer named Manas Efendi.[56]

It was deficient because of the increase in railway loads and damage suffered during WWI. Then it was sold by the Anatolian Railway Company to the Province. However, the province was not able to complete the project and it was taken over by Nafia (Figure 6.24).

The bridge was dismantled and brought to the new location. During transportation, the sections were mixed up causing difficulty in reconstruction.

The Eastern abutment at the new location on the Vezirhan side was resting on rock and was constructed in dry conditions. The abutment foundation at the western side was constructed with timber sheet piling with a depth of 2.4 m below the lowest water level and resting on 28 timber piles (Figure 6.25).

The base of the pier started to be excavated using a caisson; however, since the caisson could not be lowered after 3 m deep, further excavation was undertaken with the timber sheet piles. The peripheral sheet piles allowed the foundation to go deeper reaching 4 m. The foundation of the pier was built on top of 30 timber piles driven from the base of the excavated caisson.

While the new bridge deck floor was planned to be made of concrete with protection layers, it was found that the metal was deteriorated resulting in insufficient

Figure 6.24 Vezirhan bridge while in use as railway bridge, photograph by Guillaume Gustave Berggren, titled "view of the bridge between Mekece and Osmaneli stations over the Sakarya River, c. 1893" (İstanbul University Library - Rare Works Collection).

Figure 6.25 Vezirhan bridge (KGM album 1988).

carrying capacity. This conclusion was made as a result of tests carried out in the Laboratory of the Engineering School. The allowable stress was limited to 800 kg/cm^2 and the intended decking was replaced with timber to reduce the weight (Figure 6.26).

The tender for the bridge relocation was advertised in the newspaper on 04 August 1930[57] and the project was awarded to engineer Vahit with a contract dated 06 September 1930. It was completed on 01 October 1932.

Figure 6.26 Vezirhan bridge tender announcement (Vakit Newspaper).

MTL-04: Borçka bridge

This structure was the longest span in Turkey when constructed in 1935 with a clear span of 113 m.

It is in the Black sea region on the Hopa-Kars road, 36 km inland from the Hopa coastline and crossing over the Çoruh River. The rise of the arch is 15.75 m from the support level. The bridge is made of steel and the deck was timber with a 4.5 m width for the road.

Although a previous bridge was built at this location by the Russians, it collapsed as a result of scouring by the river, which has a steep gradient and heavy flow. After the collapse, the transportation on this vital route was provided by a temporary cable raft service.

The bridge is a Langer beam system. In a Langer arch bridge, the arch rib is thin, but the girders are deep. Therefore, it is assumed that the arch rib only resists compression forces, whilst the girders take both bending moment and axial tension. It was invented in 1883 by an engineer named Josef Langer. In this system, the main structural steel beams, which are at the edges of the deck, have constant height and are stiffened by the arch above.

It is also called Langer Beam in Europe and examples with vertical hangers are rare.

The existing abutments were utilized for the new bridge and these appeared to be in fair condition. One was standing on a solid caisson and it was found that the other was partially damaged from scour. Therefore, the superstructure was kept as light as possible for existing abutments, and a steel bridge with a timber deck was chosen for this purpose. In addition, high strength steel was used to further reduce the weight of the structure.

The major steel parts of the bridge weighed 240 tons and were made by the Skoda factory for 68,400 lira. However, since the amount of 26,000 lira was custom duties and land entrance tax, the bridge itself cost 42,500 lira. In addition, 7,200 lira was spent on the repair of existing abutments and the timber decking.

A letter was found relating to the customs duty exemption for the import of the material to be used for the assembly of the bridge. The letter also names Karl Anders as the responsible agent for the project (Figure 6.27).[58]

Feyzi Akkaya[59] was sent to Budapest for his internship to work on a bridge construction at "Boraros Square" by Nafia.[60] This bridge was Horthy Miklós (Petőfi) Bridge[61] and as we learned from Akkaya, Nafia also appointed him for the role of receiving the material of the Borçka steel bridge. Akkaya undertook research about the specified high strength steel, which was new at that time, and related technical standards and testing requirements for the handover. He collected the German norms and specifications and

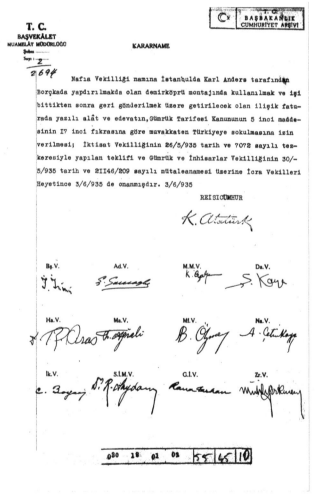

Figure 6.27 Letter for the customs duty exemption of the Borçka bridge (BCA 55-45-10).

hired a German tutor for himself. Unfortunately, the fabrication of the Miklos bridge was put "on hold" and Borçka bridge material was not ready for him[62]:

> ... the Skoda factories in Pilzen, Czechoslovakia, wrote to me, that they would invite me when the material was ready...
>
> At the end of March 1934, when the "Skoda" factories had not yet invited me and the Horthy Miklós (Petőfi) Bridge construction site in Tuna had not yet started work, I left Hungary by planting one last Tokay wine at the border.

A traveller in 1935 came across the engineers working in the bridge construction (Figure 6.28)[63]:

> ...He ...explained. The Ministry of Public Works was building a bridge over the Çoruh and he and another Czech engineer were in charge of the construction which they carried out for the Skoda-works. It was a first-class piece of engineering, a single unsupported span of about 370feet [112.77 m] in length stretched from one bank to the other, beautiful in the simplicity of its bold line. The men were now busily painting it with red lead, in a few weeks the work would be completed after altogether nine months...

Some information is provided by a resident named Mehmet,[64] reporting about Osman Arslantürk, who had carried sandbags for the test loading of the bridge when he was 12–13 years of age, as:

> After the construction finished, the bridge had been loaded with sandbags which were carried by youths of a certain age. Then the load of the sand was certain and the load-carrying capacity of the bridge revealed and hand-over was complete... the money to sand carriers had been paid by Contractor...

Figure 6.28 Borçka bridge (author's collection).

Çoruh nehri üzerine kurulan büyük bir köprü

Hopa (Husuî) — Hopaya 37 kilomet-roluk bir şose ile bağlı olan Borçka köp-rüsü bitmek üzeredir. Bu köprü, köprü-cülük tekniği ve iktisadi bakımdan çok ehemmiyetlidir. Bu köprü Çoruh nehri-nin iki kıyısına tutturulmuş asma bir köprüdür. Uzunluğu 103 metro, ağırlığı 250 tondur.

Şark vilâyetlerimizin denizle olan mi nasebatı şimdiye kadar yalnız Trabzon iskelesine inhisar ederken, bu köprü sa-yesinde bu münasebet Hopaya da çevri-lecektir. Hopa ile Kars, Ağrı, Ardahan, Artvin arasında sıkı münasebetler başlı-yacaktır. Gönderdiğim resim Borçka köprüsünü göstermektedir.

Figure 6.29 Newspaper article announcing the near completion of the bridge (Cumhuriyet 03 July 1935).

Progress for the bridge construction was reported in a newspaper report titled[65]: "A large bridge built over the Çoruh river" and informed that the eastern region will be connected to sea transport via Hopa as an alternative to the only existing connection at that time at Trabzon (Figure 6.29).

Its temporary acceptance was made in August 1935, and it was opened with a ceremony on 23 July 1935 by Prime Minister İsmet İnönü, who was on an examination trip in the eastern provinces.

MTL-05: Kemah bridge

The first major suspension bridge in Turkey!

This bridge replaced the old timber bridge[66] and connected Adatepe and Ilıç regions crossing over Fırat River; today it connects Bağıştaş village to Kayacık village.

In 1933, the project was tendered as a single span concrete arch bridge. However, the location was very remote and no candidates presented for the tender, resulting in the project not being implemented.

Considering such difficulties and also the depth of the gorge, a metal suspension bridge system was then considered because of its light weight and the advantage of not requiring scaffolding. Even though a suspension bridge system was generally applied to longer spans, it was the most feasible solution for this location with a 53 m long span.

Since heavy loading is not common in this region, the structure was designed only for a small truck of 9tons together with a weight of 100 kg/m² or alternatively with a 250 kg/m² without a truck.

The original decking of the bridge was timber. It appears that at a later stage, steel plates have been added to increase comfort and improve the distribution of the load.

Cables: The bridge was carried by a total of four main cables with 2 of 40 mm on each side, and a total of 98 m in length. Each cable has 61 strands, each 4.5 mm diameter. The cables pass over the 8.3 m high towers and are anchored with concrete blocks behind the bridge. These cables, with a breaking stress of 160 kg/mm², operate with a 4.04 safety factor under the combined load of self-weight and maximum live load of 70.40 tons.

As observed from photographs of the bridge, two cables were later changed to a single cable.

The span is 52.8 m long between the embankments. The truss has 12 bays with a length of 4.4 m and a depth of 1.7 m. The distance between the truss beams, i.e. the width of the bridge is 4 m (Figure 6.30).

The bridge is in the vicinity of Sivas-Erzincan railway line which was under construction at the same time. Therefore, photographs of the bridge found in Nafia railway albums[67] and STFA archive, which were constructing railway bridges in the area (Figure 6.31).

Figure 6.30 Kemah bridge partly seen with the timber bridge in the foreground (STFA archive).

*Fırat nehri üzerinde Kemah kasabası yanında
yol köprüsü. Uzunluğu53 metre. İkmal tarihi ; 29.11.937*

Figure 6.31 Kemah bridge from Nafia Railway album dated 1937 (İBB Atatürk Library).

The construction of the Kemah Bridge was awarded to a contractor named Raşit Börekçi,[68] who also built the Ilıç bridge in the same region. The project was completed in 12 months with the contract signed on 16 December 1935. The tower, cables and other metal parts of the bridge were altogether in the order of 55tons.

The construction of the bridge was completed on 28 October 1937.

MTL-06: Ilıç bridge

This bridge was built to replace a timber structure on the road connecting Erzincan to Malatya and Elazığ via Kemaliye and crossing over the Fırat (Euphrates) River. The timber bridge was resting on the remains of a former stone bridge.[69]

The steel bridge had a three-span arrangement of 12.40, 36 and 12.40 m. The choice of truss steel bridge type arose from the remoteness of the region. One of the piers of the bridge was indeed a large rock (Figure 6.32).

Project calculations were based on the parameters used in the Kemah Bridge for live loads. The metal weight was determined as 39.5 tons and they were manufactured in Germany.

This bridge, along with the Kemah bridge, was awarded to the same contractor, Raşit Börekçi and also finished at the same time in 1937.

As of today, this structure is no longer standing. The superstructure of the collapsed bridge is visible in the river in a photograph from 2015.

*Fırat nehri üzerinde Iliç istasiyonu karşısında
yol köprüsü. Uzunluğu 68 metre. İkmal tarihi : 29.11.937*

Figure 6.32 Ilıç bridge from Nafia railway album dated 1937 (İBB Atatürk Library).

MTL-07: Gazi (Atatürk[70]/Unkapanı) bridge

Gazi Bridge is the longest standing bridge of the numerous previous bridges crossing the Golden Horn. It is in the Unkapanı location of the Golden Horn and opened to traffic in 1939. It is currently the only floating (pontoon) bridge in Turkey.

Because of its importance, it was given the name "Gazi" meaning 'veteran', in honour of the nation's leader, Atatürk. This same importance led the bridge design to face intense scrutiny and endless debates about whether the bridge should have oriental architectural features to reflect the nation's origins or instead, exhibit a modern appearance to emphasize the new republic. These debates nearly created a revolt among local authorities. Despite, or perhaps as a result of, so much attention and emphasis, in the end, a bridge with no pleasing features was the result (Figure 6.33).

The story started in 1927 with a committee formed for the realization of the project. The committee members were renowned bridge engineers of the country[71]:

Commission members are listed as[72] (1) Fuat Bey: The ex-director of the Municipal Department of Technical Services, (2) M. Fuat Bey: Director of Technical Services of the State Railways, (3) Engineer İrfan Bey and Fikri (Santur) Bey.

The committee studied a variety of options for the bridge with permanent piers and with cables to avoid a pontoon bridge. Unfortunately, many constraints disqualified these options.

Figure 6.33 **Gazi (Atatürk/Unkapanı) bridge (author).**

The primary constraint was the difficulty of constructing a pier in the harbour. At this crossing, the Golden Horn is approximately 450 m wide and around 30–40 m deep in the middle. The Harbour bed consists of muddy soft soil of 20–30 m depth. Therefore, any fixed pier would have needed at least 65 m in depth before they find a reliable level to carry the bridge load: "technology of 1927 was not capable of achieving this. Even so, the municipality was not able to afford the expected cost "[73]

Another option was a large cable-stayed bridge needing 48 m navigational height clearance, which would require the bridge to have about 1400 m total length. In addition, the cabled bridge would not have a direct connection to the Golden Horn region. The span would extend from Süleymaniye to Şişhane and block the view of many landmark mosques in the vicinity and dominate the city silhouette.

Another point of consideration was that, the Golden Horn is a natural harbour and wind and waves are not expected to be extreme. In conclusion, a floating type was chosen to best suit the conditions and constraints.

The committee determined the conditions and a tender opened for the bridge including a design competition. Three firms, one from Germany and two from France, submitted proposals. A French firm won the competition and also had the lowest price at 4 million TL.

Funding was being raised from the toll of Galata Bridge and at that time, the collected amount reached only 300.000 TL. In consequence, the municipality did not want to proceed with any of the offers.

In 1929, the municipality approached the International Metal Bridge Experts Association in Belgium, who advised some experts in the field from amongst them. An engineer named Gaston Pigeaud was contracted for the preliminary design of the bridge.

Pigeaud Design, 1930:

Gaston Pigeaud (1864–1950)[74] was the vice manager in Ecole de Ponts et Chaussées and inspector in the Ministry of Public Works of France. His visit to İstanbul was announced on 08 May 1929[75] news as he was arriving the same day and 14 May 1929[76] news announced his ongoing investigations and visit to the Municipality.

Later, on 23 August 1929,[77] it was reported that Pigeaud was requesting further information regarding the allocation of the bridge deck and the use of tramways from the Municipality.

The progress of the bridge project was regularly reported in news articles. 06 January 1930 dated news[78] portrayed Pigeaud with scissors since the existing bridge at the location was planned to be separated into two pieces and each used for crossing narrower spans further upstream of the Golden Horn.

Pigeaud designed a pontoon bridge and submitted the project in 1930. This design drawing is found in KGM archives.[79] The design had to be sent to İstanbul twice as the first copy never arrived.

When the design sent from France was lost on its way to Constantinople, this made the news headlines with the title "wanted" on 28 February 1930.[80] It concerned the authorities, who were worried that the design might have been stolen by competitors. Pigeaud declared that the two copies of the design have been sent 1.5 months ago from Paris with the Paris-Marseille train. Then, the municipality requested a third copy (Figure 6.34).[81]

In the 2 April 1930 dated article, it was announced that the design drawings had arrived via airmail and that the review process had started.[82] The engineers involved in the review process were: Technical services Ziya (Erdem) Bey, Supervisor Burhanettin Bey and Chief Bridge Engineer Galip (Alnar) Bey.[83]

Alnar also visited Paris on 02 October 1929 to discuss design issues[84] and went to Berlin on 20 March 1937 for the handover of the bridge, where Pigeaud was also present.[85]

The design was also reviewed by various commissions of many experts and various reports were prepared.[86] The foreign experts were: Andre Ficenisky professor of bridges in the Engineering School of Warsaw, (Louis) Grelot, chief engineer in the iron bridges department of the French Ministry of Public Works and Ruttimann, a Swiss engineer who specialized in metal bridges.

The Turkish experts were from the School of Engineering: Mehmet Fikri Bey (Santur), Mustafa Hukuki Bey (Professor of mechanical engineering) and Burhanettin Berken (Professor of masonry bridges), İrfan Bey Department of State Railroads (Chief of the bridge department), Emin Bey (Engineer in the bridge department), Director Ziya Bey and ex-director Fuat Bey from the Municipal Department of Technical Services (Figure 6.35).

Endless discussions took place between authorities, mainly the Municipality and Ministry, in regards to the technical parameters, design codes and architectural features included in the proposal of Pigeaud.

As stated by Örmecioğlu[87]:

... the main discussion was based on the facade design of the proposal. The facade proposal was clearly consisting of orientalist architectural forms and ornaments. This design, which was formed in repetition of the non-structural elements reminding Ottoman architecture such as basket-handle and pointed arches in a rhythm, was unacceptable under early republican cultural politics...

Figure 6.34 Newspapers as mentioned in the text in order with the date. from left to right 1. Unkapanı bridge will be cut into Two pieces! (6 January 1930) 2. Wanted! Where is Mr. Pigeaud's Project? (28 February 1930) 3. Gazi Bridge. Tender will be opened for the Bridge (3 March 1930) 4. Gazi Bridge Project. New Bridge will have the same length and width as the Karaköy Bridge (2 April 1930)

Figure 6.35 1930 dated Pigeaud design for Gazi bridge from KGM achieves.

While the ministry was criticizing the proposed details, the municipality was defending the proposal. The main standpoint of the ministry was the design of the bridge as: '…not compatible with the modern taste and contemporary technological advance in terms of both facade design and steel structure'[88]

> …while the general view of the bridge complies with the taste of the imperial times and suits with 20 years earlier conceptions in other countries, it is not able to satisfy the taste of the republican generations who aim at working with modern and latest methods, and building the country.[89]

On the other hand, the Municipality defended the forms as being functional elements:

> …the basket-handle arches framing the bridge underneath, were aimed for ships and steamboats crossing Golden Horn day and night to help perfectly align with the spans, and easily pass under[90]

Discussions went beyond the structure to and into the embracing of the architectural style of the nation:

> … the municipality stood for the pointed arch form with a nationalistic defence as a noble form of Turkish architecture descended from great Turkish masters

such as Sinan, Kasım, and Hayrettin but not of degenerated Ottoman architecture created by foreigners in the decline of the Empire. Conversely, the Ministry considered the bridges as culturally and politically constructed symbols of technology and modernity, which had to be in accord with "the taste of the nation on its way to modernism".[91]

These long debates continued until 1933. Although the municipality insisted on defending the design as not being orientalist but rather functional; the ministry would not approve the design of such an important bridge unless 'it was...appropriate for İstanbul, the Republic and the [Turkish] reforms'.[92]

The law authorizing the municipality to make independent approval of designs and proposals of local projects in İstanbul was revised in 1935, under the pressure of the ministry. The new law limited the decision-making powers of the municipality and defined the Ministry of Public Works as "the component authority on approval of the designs and proposals of the piers and bridges projects in İstanbul costing more than fifty thousand Turkish Liras".[93]

The Pigeaud project was revised and put on tender in 1935.

The debates, (or arguments), even continued through the construction of the bridge. Alnar described the bridge in a newspaper column in 1939.[94] His article gave a considerable comparison with the 1912 Galata bridge.[95] He compares their main features such as span, height, width, pontoons and especially the style, etc. For example, he explained that the main difference between them is the connection systems, stating that the Gazi Bridge with fewer connections between parts would reduce the dynamic impacts of vehicles, and declares Gazi bridge as the "the biggest, the longest and perfect".

According to Alnar: Galata bridge balustrades are adorned with the motifs, which are used in parapets and windows of the mosque near the western end of the bridge,[96] whereas the Gazi bridge did not need any "motif development". He considered its "Modern style" was fit for purpose.

Further Proposals: Although the project was already commissioned, alternative projects continued to be published in the news. Even though most were not formal and only made it to the news, some are worth mentioning. The first one was made by a man named "Cim Türk", the second is a proposal from Architect Ernst Egli, the third is a replica of Tower Bridge in London and the last a proposal from American Engineer Waddle.

Mr. Cim Türk's proposal, found in article dated 30 June 1930,[97] consists of three stories with a huge pier in the middle and detailed in the newspaper as follows:

> A person advises the bridge between Süleymaniye and Tunnel with three stories:
> ... bridges are the life arterials rather than the jewels of İstanbul ... The bridge will be abutting to Süleymaniye at İstanbul side and Tunnel square next to Galata tower on the other side. The bridge will be approximately 1200 m in length, 60 m in height and 4 m in width, and the first floor will be allocated to

pedestrians, the second floor for vehicle transportation, and the third floor for trams with limitations.

The tower in the middle and bridge ends should be made in such a way that the width of the bridge can be widened if needed. The tower on the landside will also serve as an entrance, and the people will go up and down with the elevator. The height of each floor will be 3 m and shops of suitable size will be built inside. In the middle of the bridge, on top of the columns that rest on the tower's pillar, a magnificent light will enlighten the 'Golden Horn' with an electric sun with a power equal to 1 million candle lights...The old bridge should be taken to the shipyard to recycle.

The proposal by Egli was a suspension bridge from Süleymaniye to Galata Tower. Ernst Arnold Egli was in Turkey during 1927–1940 as a teacher at the School of Fine Arts and was also involved in public and private projects. On 5 July 1932, a Cumhuriyet newspaper shows the location of the proposed bridge in between the existing bridges. The news article mentions that one of Egli's students also drew plans for the bridge (Figure 6.36).

The third proposal, made by a British Firm, was a replica of Tower Bridge in London. A 22 June 1933 dated article[98] mentions that the project details were submitted. The cost of the suspension bridge would be around 6 million TL; however, this option would not need as frequent repair works as a pontoon bridge option.

Figure 6.36 Left: Cim Türk's proposal, right: tower bridge proposal.

A Vertical Lift Bridge design proposal in 1934 was made by a prominent Bridge Engineer Waddle (1854–1938). Waddle had engineered many bridges and his most important contribution was the development of the steam-powered vertical-lift bridge for large crossings.

His first design was in 1893 for Chicago's South Halsted Street Lift-Bridge over the Chicago River and the last bridge was Lower Hack Lift in New Jersey, very similar to the proposal made for the Golden Horn.

Waddell, together with his partner of that time named Hardesty, proposed a vertical lift bridge and heavily criticized the floating bridge option referring to it as the use of old technology of two or more decades earlier, stating:

"It is obvious that in the admirable modernization plans, which are being efficiently pursued by the exceedingly foresighted and able leaders of the progressive Turkish republic, the most advanced scientific types of public constructions should be adopted; and it is to be hoped that American engineers may be given an opportunity to contribute thereto".[99]

Waddell prepared proposals for two types of vertical lifts; the first option had a common type of vertical lift bridge with lattice towers, which carried the counterweight and machinery, with a lifting span in between. The second option is different as it has only part of the span lifted and it is not supported on piers. The proposed towers had an enlarged base integrated with the truss superstructure (Figure 6.37).

It was during this period that city planner Henri Prost[100] was invited to prepare the master plan of İstanbul by the government in 1935.

Prost's plan imposed a restriction on the maximum height of structures in the Historical Peninsula and Bosphorus to protect the city profile. The height limitation is defined by Prost as: 'The height of buildings to be constructed in areas that are 40 m or more above sea level is not to exceed 12 m in height, and that construction on lower levels is not to exceed those heights at the 40 m level.'[101]

This height restriction made the cabled bridge option not complaint to provide navigational clearance as it would exceed the limits specified by urban planning.

Financing the Project: Since 1928, the Municipality had been collecting taxes from mass and private transportation for the maintenance of Golden Horn bridges and the construction of the new Gazi Bridge. This tax was in the form of toll and was approved by Law No.1223. This law was modified in 1933 with an increase for vehicles, whilst pedestrian crossings were made free of charge. Ceasing the pedestrian toll which had started a century ago was celebrated by the city's population like a carnival!

Pigeaud's project was put on tender on 17 October 1935. Die Maschinenfabrik Augsburg-Nürnberg AG (MAN) company was awarded with the contract. The company was also the contractor of the 1912 Galata bridge which served until 1992. The winning consortium consisted of four forerunner firms of the German steel industry[102] and MAN was the managing director. Hugo Herman and Galip Fescioğlu[103] were the agent middlemen working for MAN in İstanbul. Chief engineers were K. Karner and Dipl.-Ing. Arnold Paul.

Sadık Diri and Ferruh Atav Co. was the subcontractor for the construction of pile foundations. The project was realized under the responsibility of both the ministry and the municipality. The control engineers were Necati Turfan and Sadi Cimilli from the municipality. An Engineer named Balaj also attended final controls on behalf of the ministry.

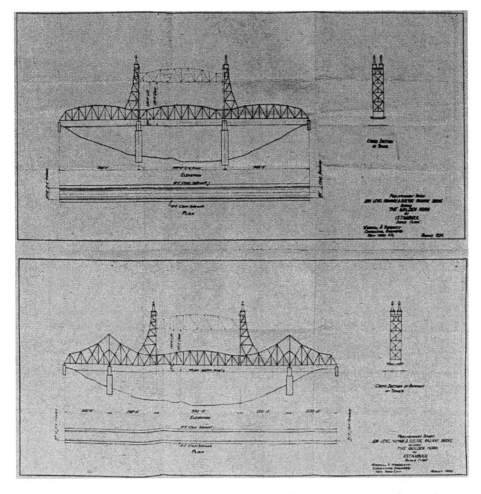

Figure 6.37 Drawings of the two lift-bridge proposals of Waddell-Hardesty Co. on August 1934.

A ceremony for the commencement of foundation works at abutments was held on 29 August 1936 with the attendance of a huge crowd and flags were hung across the width of the road. Governor Muhiddin Üstündağ ceremonially placed coins underneath the foundation, which were covered with concrete. The next day, 30 August 1936, photographs from the ceremony were printed on the cover page of most of the newspapers.[104] (Figure 6.38)

Then a declaration was written stating:

> This bridge, whose construction costs were provided with money obtained by the people of İstanbul, was attributed to the Great Leader Atatürk; the foundation has been laid on Saturday the twenty-ninth of August one thousand nine hundred and thirty-six.

Figure 6.38 The ceremony for the start of foundation works, news collected from Akşam, Son Posta, Cumhuriyet and Tan Newspaper (in order from left to right) all dated 30 August 1936.

The leaflet was signed by Deputy Internal Affairs Şükrü Kaya, Minister of Monopolies Ali Rana (Tarhan), Ms. Hakkiye (Koray) Emin,[105] Parliamentarian Münir, the Former Minister of Public Works, the Director of the Science

Delegation and the Director of Bridges, and then placed in the foundation with the coins.

The construction progress can easily be traced through news articles: 20 November 1936[106] dated news with heading as "Gazi Bridge Construction" reported that:

..the black cement, ordered from Germany, hasn't arrived and the concreting hasn't yet started. This black cement was ordered 2 months ago, but still has not arrived. Therefore, only some piling work on the İstanbul [west] side has taken place, and with the rest of the foundation work put on hold, as there is no cement and no work has started on the Beyoğlu side... The contractor named Hugo Herman, who undertook the construction of the bridge, had ordered this cement from Nafia.

In news dated 10 March 1937[107] the heading declares "Gazi Bridge Construction Progressing":

The pontoons ordered from Germany will arrive in four months. The test piles prepared for the Gazi bridge are about to be put in their place.

...The preparations for the construction of the Gazi bridge continue feverishly. Bridge engineer Galip [Alnar] went to Germany and he will also be in contact with the expert Piju [Pigeaud]. Galip also took preliminary designs with him.

Some of the pontoons being built in Germany are finished. These pontoons are numbered in pieces and packaged and will soon be sent to our city to be opened and assembled in the Balat workshop. Currently, the test piles are being driven at the Unkapanı side. The concrete piles driven recently to 15 m deep did not reach stable ground since the soil is very soft on this side, and 18 m long new piles are being poured at the moment. It is understood that such depth will be sufficient. After the results are obtained from the first test piles, 180[108] piles will be placed on the Unkapanı side of the bridge. Since Azapkapı side ground profile is better, 100 piles will be enough...

An article from 12 March 1938[109] provided further information for the construction of the pontoons:

The assembly of pontoons for Gazi Bridge is progressing well in Balat. Yesterday one more pontoon was launched in the sea. Pontoons will be 24 in total, and together with the spares, there will be 26.[110] Currently, 200 workers are on duty. Each pontoon is assembled in approximately 20 days. One of the pontoons is set in its place and assembly of the pier columns and the parapets commenced.

23 May 1939[111] dated news announced the handover of the bridge from contractor to authority expected on 20 August 1939. Construction was completed and pedestrian handrails were mounted, sidewalks were asphalted and the rest of the deck was laid with the waterproof membrane ready to receive the final layer. The final layer was to be a timber parquet, ordered from France, chosen for its durability and lightness. The rails were also installed on the bridge as per the contract; however, the final decision for the tramway to operate was not concluded yet.

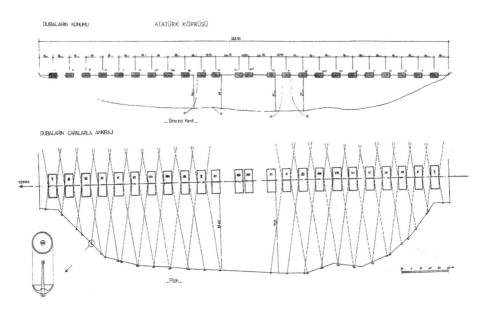

Figure 6.39 Gazi (Atatürk/Unkapanı) bridge sketch by İsmet İlter (İlter 1973.)

The targeted handover could possibly be delayed because of the late arrival of the imported timber parquet.

The Bridge: The total length of the bridge from bank to bank is 453.50 m. The two anchored parts, referred to as regular spans, rest on ten pontoons, whilst the movable middle part consists of four pontoons. This configuration provides approximately 8.5 m wide clear passage of small boats at the regular spans and 15.5 m wide in the two middle spans.[112] The end spans are used for the transition from land to the first pontoon which has a pivot connection against the abutment, which can thus compensate for the variable level of the water on which the pontoons float (Figure 6.39).

The bridge consists of five structural parts: two end spans, each 19 m long, two anchored parts, each 169.5 m long and the middle movable part of 76 m length. Similar dimensions are given as 19, 170 and 70 m by Alnar. Alnar also informs that the bridge length has been extended to 477 m as per the revisions of Prost.

The bridge has 24 spans and 24 pontoons as the two middle pontoons join directly. Each of these was 25 m long and 9 m wide 3.4 and 3.8 m in depth. They were made of St52 steel, whilst the superstructure is made of St37 steel. All of the pontoons were made the same size to be interchangeable with the spares and had 16 separated compartments inside them to avoid the risk of having the entire unit being submerged if a water leak leads the interior being filled with water. The pontoons are detailed following shipbuilding principles. For example, plates are joined by overlapping and notching and the outer plates are finished with mortice edges.

Anchoring of the structure is not done directly through the pontoons. Instead, the steel castings in which the anchor chains are hung, are connected to columns at their base. This arrangement has been chosen to facilitate the replacement of the pontoons for maintenance.

The width of the bridge is 25 m including 4.5 m wide pedestrian sidewalks on both sides.[113] Seven longitudinal beams are spaced at 3.33 m and topped with a 16 cm reinforced concrete slab.

A tugboat was used for opening the movable section which took about 7 minutes. The opening was scheduled once a day in the early morning hours.

As a requirement of acceptance, load testing of the bridge was carried out measuring the amount of sinking of the pontoons. Loading was performed with 12 loaded trucks, each weighing around 8 t, which were arranged in two rows of six next to each other and then in four rows of three next to each other.

The opening of the Bridge was made with a formal ceremony held on 29 October 1939. In his speech during the inauguration, Lütfi Kırdar, the Governor and Mayor of the city, announced that the bridge was to be named "Gazi" as a "symbol of the gratitude [of the public] to the eternal saviour". He proceeded to cut the ribbon after his speech and then invited the public to enjoy a dedicated buffet prepared by the contractor.

Survival of the Bridge: The roadway had 14 cm thick timber parquetry over a concrete slab with protection layers in between as per the original plans. Alnar advocates this choice to reduce the weight and the noise of the ongoing traffic. The bridge was completed with timber before its opening in 1939. Less than a week after the bridge was opened, this pavement parquetry had crumbled. A columnist writing about the event finishes his article[114] as:

> Miserable Cisr-i Cedid[115] who cost fortunes. I retrieve the praises, I once said unconsciously, from these columns.[116] I hope it will not be our generation's children's destiny to see that one day you too are crushed by a strong storm like your old grandfather. Goodbye for now, miserable bridge that humped at a young age...

comparing the crushed tiles to the humped back of an elderly person.

After undertaking the necessary repair works, the bridge opened to vehicle traffic again on 20 February 1940.

The Golden Horn pontoon bridges have been a considerable work for Municipality to maintain and keep safe. There have been occasions where water leaked into the pontoons and had to be discharged urgently to prevent the sinking of the sections of the bridge.

Atatürk bridge still serves road traffic.

MTL-08: Ceyhan bridge

Ceyhan Bridge is an arch type bridge with an 80 m main span. It is in the district of Ceyhan in Adana province and crosses over the Ceyhan river. The bridge deck is connected with ten hangers and the connections are riveted.

Ceyhan bridge superstructure was ordered from Die Maschinenfabrik Augsburg-Nürnberg AG (MAN) and the substructure was constructed by SAHA and Ziya Bağlar (Figure 6.40).

Robert W. Hunt carried out the testing of the bridge for its handover.

Figure 6.40 Ceyhan bridge (KGM Album 1988).

Archive documents relating to the payment and progress of works provide a time-line history of the bridge construction. As per the letter[117] written by Nafia, the sub-structure works were undertaken by contractors Sadık Diri and Halit Köprücü,[118] commencing with a contract signed on 06 May 1937 with a duration of 12 months which was due for completion on 06 May 1938.

However, the work duration extended by another 182 working days because of the challenges encountered during the construction. Groundwater demonstrated an excess of sulphate content that might damage the concrete, the geotechnical conditions were found weaker than predicted (or investigated) and floods occurred during the construction. As a precaution for sulphate attack on concrete parts, sulphate resistant cement was used in the concrete mix and the cement amount in the concrete mix was increased.

Nafia agreed to the extension of the work duration and also increased the payment corresponding to roughly 45% of the tender value to the contractor (SAHA) for the additions and project revisions as learned from the letter.[119]

By the end of 1939, the substructure was completed and ready for the installation of the superstructure; however, as per the original contract, the company MAN contracted for the superstructure works had sent their team for the works according to the agreed contract date which was the 08 December 1938. The team for assembly brought the bridge but had to leave as they could not commence the assembly.

There is also another letter[120] requesting permission for the entrance of the assembly tools and machinery for the assembly of the Ceyhan Bridge. Tools and machinery would be returned to the country of origin after the works have been completed. This letter is dated 05 October 1938 and was signed by Atatürk (Figure 6.41).

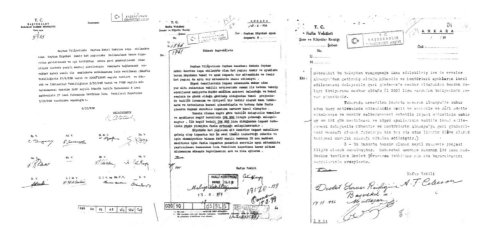

Figure 6.41 Various letters for the decisions regarding the challenges faced during the construction of Ceyhan Bridge, from left to right: 05 October 1938, 10 February 1939 and 26 December 1942.

MAN was later requested to resume the assembly work in May 1940. The delay was to avoid the high risk of flood during the winter season. However, MAN could not start the work until November 1940 and could only send one expert because of the challenging conditions of the war. The works were finally completed on 12 October 1941.

MTL-09 and MTL-10: Sırzı and Kozluk bridges

Sırzı and Kozluk bridges are both on the road from Malatya to Erzincan and built during 1942–1943. Both bridges were built by an engineer named David Parker[121] and the steel was bought from the Witkowitzer Bergbau firm of Czechoslovakia as indicated in the KGM list.

Sırzı was a 40 m long single span deck arch bridge with a braced spandrel. It probably remains submerged under the Keban Dam lake constructed between 1966 and 1974. For the Kozluk bridge, only a street view is available in Google maps. It shows a truss bridge with riveted connections (Figure 6.42).

The engineer Dr David Parker acted as the representative engineer of the contractor company for the construction of Pertek Bridge. During the 1940s, he also undertook a variety of steel bridge projects such as Kozluk and Sırzı Bridges. However, he had to close his firm with a loss due to difficult pre-war conditions.

Örmecioğlu provides further information[122]: "The transportation fees and insurances increased under high risks of war, this was added to the total cost of the bridges, and contractors were stranded as in the cases of Kozluk and Sırzı Bridges".

Figure 6.42 Sırzı bridge shown during the road construction works of Keban Dam.

MTL-11: Kirazlık bridge

This bridge is in İzmit/Kocaeli on the road leading to Değirmendere. It is included in this book as it is an unusual case of a recycled bridge.

From Örmecioğlu[123]: "... 12 m spanning Kirazdere Bridge was constructed with the steel elements donated by the General [Military] staff for the construction of militarily important road bridges. The steel beams were bought from England for gun emplacements during WWII..."

The bridge can be seen on Google maps as four spans of 12 m each and 48 m total length. There is another road bridge built next to it.

MTL-12: Mameki bridge

Mameki bridge was on the road from Elazığ to Erzincan crossing over the Munzur River which joins Fırat. According to KGM documents, its span length was 56 m. The bridge is not found and probably is submerged in the later constructed Keban Dam lake.

As stated by Örmecioğlu:

'Until the construction of Mameki Bridge in 1946, the Ministry preferred to bid to foreign companies through agent middlemen instead of inviting offers directly from the producer companies. However, this method caused an increase in costs because of the commission fees of the agents and caused some problems because of lack of direct contact. Hence, the ministry opened an international bid and directly contacted the producer companies for Mameki Bridge with a cabinet decision'

The decision letter [124] dated 20 May 1946 relates to the steel material needed for the 56 m long Mameki bridge, which was intended to be built at the 90th km of the Elazığ-Erzincan road:

'...As per the law first article of the law numbered 4097, it has been decided to purchase the mentioned bridge material with bargain by directly contacting the companies or their legal representatives without the intervention of commissioners or brokers, not seeking for security bond and delay compensation, and taking into account any increase or decrease in prices. The payment will be done as 20% of the contract value for the order and transportation costs and other expenses as advance payment after the contract is made, 70% when the material is ready for shipment and 10% when the material arrives to our country.'

MTL-13: Genç bridge

Dated in 1952, this structure is in the Genç district of Bingöl and on the main road of Bingöl to Diyarbakır. It crosses the Murat River, which is a major tributary of the Fırat (Euphrates), with three spans, each of 55 m, reaching a total length of 165 m.

The bridge has been registered in the highway records as provided by Railways and Ports Authority.

Örmecioğlu[125] mentions the bridge as: "In 1948, instead of consulting a company, the Department of State Railroads personally undertook the construction of Genç Bridge and completed in 1952. The steel parts were fabricated in Germany".

When constructed in 1952, the bridge positively impacted the region and inhabitants providing a permanent crossing over the Murat River, which was previously undertaken by the use of rafts and boats.

Today, this structure remains standing, but as a "bridge to nowhere", because it is not connected to the other side of the river. From photographs, it appears that the road embankment at one end has completely collapsed. This is probably the result of scour at abutment and lateral shift of the riverbank (Figure 6.43).

In 2014, the General Directorate of Highways (KGM) decided to demolish the bridge. However, this plan was abandoned because of the strong local opposition.

Townspeople consider the bridge as a part of their heritage: The demolition of the bridge, which is one of the greatest historical heritages of Genç, was described as barbarism and vandalism[126]: "The Authorities who believed that there could be a civilization without culture, or being civilized without being cultured, decided to demolish this bridge to provide favour to some. KGM must stop the demolition decision!"

The people who argued that the destruction of the bridge, which is one of the symbols of Genç, should be stopped put forward many arguments such as:

It has settled in the corner of many childhood memories and it should continue to settle. KGM is trying to erase this historical bridge from public and social memory. In many provinces and regions, such historical assets are restored and taken under protection, while in Genç, history is destroyed. The historical bridge that needs to be opened to tourism is sacrificed for someone's interests and dirty games. We invite all officials, bureaucrats, politicians and environmentalists to be sensitive to prevent the destruction.

Figure 6.43 **Genç bridge (Photo: "Gencin Sesi" Local Newspaper).**

The demolition of the bridge was advertised for tender in 2014. The estimated weight of the bridge to be demolished was stated as 572.181kg of metal. The tender was to be held on 24 June 2014 at the address of the 8th Regional Directorate of Highways in Elazığ.

In a victory of its local supporters, the demolition decision was cancelled as a result of the public reactions after the demolition declaration in the press and with the initiatives of the Genç Municipality. The 8th Regional Directorate of Highways quickly suspended the demolition announcement of the Bridge, which was declared by the Public Tender Authority.

MTL-14: Bartın bridge

Bartın Bridge is Turkey's first movable bridge, if Galata and Yıldız bridges are not counted. This bascule bridge was built between 1961 and 1966. Akkaya mentions this bridge under his description of Bartın Port. He explains that in 1961, Sezai Türkeş[127] worked on this bridge one-to-one during all the stages. The design of the bridge belongs to a mechanical engineer named Mazhar Yüngül.

Presumably, a movable bascule bridge was required to provide access to port traffic.

The bridge has five spans in total with a symmetrical arrangement. The movable span of the bridge is a 15 m single leaf and 8.5 m wide. Bascule bridges are generally divided into several groups according to their mechanisms and systems. Bartın bridge

Figure 6.44 **Bartın bridge in its open state (FATEV archive).**

is a bascule bridge type characterized by a "balance beam on top". Since this type is very common in Holland, it is referred as a "Dutch type" in some reference sources (Figure 6.44).

The mechanism of the bridge is powered by an electric motor. The counterweight at the Bartın bridge is the last latitude of the upper beam placed at the back. This beam is filled with concrete to provide the required weight.

The bridge that opens as a single wing must have been designed and built quite well since it has been in service for nearly 50 years. These types of singular bridges, which are maintained by local authorities, create a real challenge as their maintenance requires special experience and equipment.

The connections on the bridge are all pin connections so that the bridge elements can rotate and open. The elements, which remain in a full vertical plane when opened, provide maximum clearance for ship passage. Suspension systems and rollers can also be seen.

The bridge and its social value are well described in an article by local historian Çetin ASMA.[128]

NOTES

1. Karaoğlu, O. (1994) XIX. Yüzyıl Osmanlı Sanayileşme Teşebbüsleri ve Zeytinburnu Demir Fabrikası'nın Kuruluşu. MA Thesis. İstanbul University.
2. Şenol, S. B. (1994) Tülomsaş; Türkiye Demiryollarında 100 Yıl. Eskişehir. 1994. Page: 55.
3. Çankaya, M. (2013) Cumhuriyet Dönemi Teknoloji Tarihi: Tarım alet ve makineleri teknolojileri, demir çelik üretim teknolojileri ve demiryolu teknolojileri. Ph.D. Ankara Üniversitesi. Bilim Tarihi Bilim Dalı. Page: 294.

4. The Mechanical and Chemical Industry Corporation (Turkish: Makina ve Kimya Endüstrisi Kurumu or MKEK), was a reorganization of government-controlled factories in Turkey that supplied the Turkish Armed Forces with military products.

5. **Selahattin Şanbaşoğlu** (1907–1995): First Metallurgical Engineer of the Republic. Was educated in Germany.
   Kiper, M. (2004) *Fabrikalar Kuran Fabrika Kardemir ve Türkiye Cumhuriyeti Demir-Çelik Öyküsü. Mühendislik Mimarlık Öyküleri-I.* IMO Publications. Ankara.

6. Somen, A. D. et al. (2015) Tarihi Çamçavuş Demir Köprüsü'nün Taşınması. 5. Tarihi Eserlerin Güçlendirilmesi ve Geleceğe Güvenle Devredilmesi Sempozyumu.

7. Photographs are taken by Vassilaki Kargopoulo (1839–1886) who was assigned as the official court photographer

8. Geyikdağı, V. N. (2011). *Foreign Investment in the Ottoman Empire: International Trade and Relations 1854–1914.* Tauris Academic Studies. London. Page: 141.

9. The title blocks of the drawings as follows in the order from left to right: BOA-1531: Jambes Namur/Theophile Finet. Bridge over Meander River dated 1914, BOA- 4164: Société Anonyme de Travaux Dyle et Bacalan. A Metal road bridge with concrete deck dated 1879, BOA-1377: Ouvraces Dart = Engineering Work (?). Typical Road Bridge projects – no date, BOA -5577: Baume Mapent 150 m Long Bridge over Gediz river - no date and BOA-2141: Vilayet de Aydın-Manisa- Soma Road bridge over Gediz River dated 1891.

10. **Heinrich August Meissner** was the chief engineer of the construction of two lines, Hijaz and Anatolian-Bagdad lines from Germany. He spoke Turkish fluently. After completing the Hijaz line, he was awarded the title of pasha in 1904 by Abdülhamit ll. In 1918 Meissner returned to Germany but he returned back in 1924 as adviser on building and maintenance of railroads in the new republic by the invitation of Atatürk. Meissner, A. H (1940) Demiryollar, Cilt 1. İTU İstanbul. Translated to Turkish by Prof. Dr. Enver Berkmen. Page: 198.

11. Peynircioğlu, H. (1951) *Köprüler - Cilt 1.* Teknik Okulu Yayınları. İstanbul.

12. The term used for Public Works was "imar", means civilization and development excluding the social content. Term later changed to "Nafia", which can be translated to utility, to refer to public assembly at the governance level. Nafia also later changed to "Bayındırlık", means prosperous, in 1935.

13. Turkish state archives can be reached at https://katalog.devletarsivleri.gov.tr. The abbreviations mean PLK: Plan, project and kroki, HRT: Map. EHT: Turkish with Arabic letters.

14. BOA 1531–1914 - Z 1531 Nehir üzerine yapılan köprü planı. (Fr.)-kesit from drawings bridge crossing Menderes river, the road project is not described.

15. **A Zores section** is a beam section designed by French engineer, Charles Ferdinand Zores, in the second half of the 19th century. It was used as surfacing on steel bridges, covering the deck girders and forming the base on which ballast, concrete or plaster for the actual deck was laid on.

16. **Macadam** is a type of road construction, pioneered by Scottish engineer John Loudon McAdam around 1820, in which single-sized crushed stone layers of small angular stones are placed in shallow lifts and compacted thoroughly.

17. BOA 4164 -Z 4164 -Demir köprü etüdü. a.g.y.tt (Fr.)

18. BOA PLK 2844 Z 2844 Gediz üzerine yapılması düşünülen köprü kesiti. (Fr.)

19. BOA-PLK-1377. Date: H 1335/M 1916: Metalik köprü ayakları ve tretuvar planı. a.g.y.tt (EHT., Fr.)

20. BOA 5737-Köprü montaj planı. (Fr.)

21. Örmecioğlu, H. T. (2010) Technology, Engineering, and Modernity in Turkey: The Case of Road Bridges between 1850 and 1960. Ph.D. thesis submitted to Middle East Technical University (METU).

22. "Mounter" term used in the list for steel workers given by KGM and "Monitor" used by Atayman and, both refer to people working in the assembly of the bridge and probably come from French 'monter' meaning assembly. Can also specifically refer to a person who works at height.

23. **Mustafa Şevki Atayman (1872–1958)** He graduated from Hendese-i Mülkiye in 1897 and appointed to Ankara and Kosova Governance as an engineer. He worked in Hedjaz

Railway between 1913 and 1918. After war, he worked in railway projects. He wrote his memoirs in a book.

Atayman, M. S. (1967) *Bir İnşaat Mühendisinin Anıları, 1897–1918.* Baha Matbaası. İstanbul.

24. Uzun, T. (2008), Geç Osmanlı-Erken Cumhuriyet Dönemi Mimarlık Pratiğinde Bilgi ve Yapım Teknolojileri Değişimi: 1906–1938 Erken Betonarme Örnekleri. YTÜ. PHd. Thesis. İstanbul

25. Lutenegger, A. J. (2013) "Historical Application of Screw-Piles and Screw-Cylinder Foundations for 19th Century Ocean Piers". International Conference on Case Histories in Geotechnical Engineering.

26. Mahan, D. H. (1873) *A Treatise on Civil Engineering.* John Wiley & Sons. New York. Page: 197.

27. Bridge drawings for this Gediz bridge found in the archives has different name as Magnesia (Aydın)-and Akhisar bridge. Possibly two bridges constructed with the same design. BOA-1400. Date: 1903: Manisa Akhisar müstakil köprü-yol planı. (Fr.)

28. Atayman, M. Ş. (1984) Page: 24

29. BOA 3426 Gediz Nehri üzerine yapılacak köprünün planları. (Fr.)

30. Peynircioğlu, H. (1951) *Köprüler - Cilt 3. Celik Köprüler.* Teknik Okulu Yayınları. İstanbul.

31. Luretenegger (2013).

32. Örmecioğlu (2010) Page: 146.

33. BCA-30-10-0-0/Muamelat Genel Müdürlüğü, 'Demiryolları İdaresi'nin İngiltere ile yapacağı satın almalarda Rubert W. Hunt Company Müessesesi'ni mübayaa ve kat'i tesellüm memurluğu ile görevlendirdiği.', Location: 151 - 68 – 18, Date: 25.02.1943.

34. **Regie (Régie) Company** was formed by a consortium of European banks and was the largest foreign investment and cooperation in the country. The capital of the company made up around 23% of total foreign direct investment in the Ottoman Empire in 1881–1914.
    Ref: Birdal, M. (2010) *The Political Economy of Ottoman Public Debt, Insolvency and European Financial Control in the Late Nineteenth Century.* Tauris Academic Studies. London.

35. Tekeli, İ. and İlkin S. (2004) *Cumhuriyetin Harcı: Modernitenin Altyapısı Oluşurken.* Bilgi Üniversitesi Yayınları. İstanbul. Page: 209.

36. Some examples: BOA-PLK- 4404: 1884 dated bridge project made for Aydın Province and PLK- 3597: 1888 dated bridge project prepared for Silifke-Karaman road by Engineer Fisbach. Both drawings are prepared only in French.

37. Martykanova, D., and Kocaman, M. (2018). *A Land of Opportunities: Foreign Engineers in the Ottoman Empire.* Philosophy of Globalisation. De Gruyter. Berlin. Pages: 237–251.

38. Atayman, M. Ş. (1984) Page: 22.

39. Kondoktör: Middle-ranking auxiliary technical staff. The title was also given to students of the engineering school who failed twice to pass to the following grade.

40. Technicians to work in assembly of bridges are called monitor or mounters.

41. Budak, A. (2019). Demir Köprü İnşasına Bir Örnek: Kırıkkale-Yahşihan Köprüsü. Akdeniz Sanat, 21. Uluslararası Ortaçağ ve Türk Dönemi Kazıları ve Sanat Tarihi Araştımaları Sempozyumu Bildirileri. Pages: 168–181.

42. BOA 3263 Yahşihan-Ankara yolunda Kızılırmak üzerine kurulacak köprünün planı. (Fr.)

43. In the archives, bridge drawings are replicated under BOA -1379: Metallic bridge plan in Kamışdere on the İzmir-Marmaris road, Aydın province. (Fr.) and BOA 2377: Plan of the Kamich Dere iron bridge between Ahiköy and Milas on the Izmir-Marmaris road. a.g.y.tt (Fr.)

44. Album (1929) *Vilayet-i Hususi İdareleri Faaliyetlerinden.* Hilal ve Cumhuriyet Matbaaları. İstanbul.

45. The text on the photograph read as: Ahi Iron Bridge on Meke?-Milas road, construction cost:3000 liras, Build date: (1)330.

46. BCA 30-18-1-2, 109 70 11 Muğla'nın Yatağan ilçesinin merkezi olan Ahiköy kasabası adının Yatağan olarak değiştirilmesi.

47. Metal bridges still survive from Ottoman Nafia period are: Çamçavuş, Kağızman, Pernavut, Yıldız Bascule Bridge, Çorlu, Yahşihan, Yatağan and Çine bridges.

48. Given as 'French Company' in Nafia report, it could mean Regie who was given concession for road project. The superstructure design drawings are not found in archives.

49. Nafia (1933) *On Senede Türkiye Nafiası 1923–1933*. Nafia Vekaleti Neşriyatı. İstanbul.
50. BOA-HRT-h-01605: Aydın-Muğla yolunun Aydın-Çine arasını gösterir harita. a.g.y.tt, Fr. (Ölçek 150000).
51. Nafia Magazine (Administrative Part) Year: 3, No:11 April 1937 Map of Aydın-Muğla road layout.
52. BOA-PLK-1664:1914 dated bridge crossing over Menderes with single span of 120m.
    Ottoman archives contain design sheets and substructure details for the bridge crossing over Menderes for Izmir-Marmaris road at Km 7 + 400 with BOA-PLK-3173: Calculations, BOA-PLK-1664: Bridge drawings and BOA-PLK-3622: Substructure details.
53. BOA-PLK-1531: Nehir üzerine yapılan köprü planı. (Fr.), BOA-PLK-3622: Marmaris yolundaki nehir üzerine yapılacak olan metal köprünün planı. a.g.y.tt (Fr.) AND BOA-PLK-3672: Menderes Nehri üzerindeki metalik köprü planı. (Fr.)
54. BCA - 030_0_18_01_02_22_59_009 Date: 17.08.1931: Antalya-Alaiye yolu üzerindeki Manavgat köprüsü inşaatının pazarlıkla yaptırılması
55. Örmecioğlu (2010) Page: 146. Reference: Unclassified documents from the State National Archives-Republican Archives, KGM Fund, Binder no: 1855.
56. Gürel, Z. (1988) Kurtuluş Savaşı'nda Demiryolculuk VIII. Belleten. C. LII. S.205. Page: 4, 5.
57. Nafia Vekaletinden (1930 August 4) Bridge Tender Announcement. Vakit Newspaper.Page: 4.
58. BCA-30-18-1-2_55_45_10_1 Date:03.06.1935 : Borçka'da yaptırılmakta olan demir köprünün montajında kullanılmak üzere montaj aleti getirilmesi.
59. **Feyzi Akkaya** (1907–2004) a prominent bridge engineer of the early Republican era. Together with **Sezai Türkeş** founded STFA Company, which was a leading engineering company in Turkey and worldwide. Their inventions and novel applications have been a vital contribution to the engineering field in Turkey. See Chapter 3 for expanded biography.
60. BCA-30-18-1-2 40 - 75 - 9 Date: 23.10.1933: Bararos Meydanı köprüsü inşaatında staj görmek üzere Yollar Umum Müdürlüğü mühendislerinden Fevzi'nin Budapeşte'ye gönderilmesi.
61. The bridge was a truss constructed between 1933 and 1937, and was named after Admiral Horthy Miklos. Today's bridge is the rebuilt one in 1952 and named after poet Sándor Petőfi. It is three span crossing over Danube with continuous truss span.
62. Akkaya, F. (1989) *Ömrümüzün Kilometre Taşları: STFA'nın Hikayesi*. Bilimsel ve Teknik Yayınları Çeviri Vakfı. İstanbul. Page: 42.
63. Linke, L. (1937) *Allah Dethroned: A Journey through Modern Turkey*. Constable & Co Ltd. London.
64. Comment provided by Mehmet Karamert at Artvin Bir Sevdadır facebook: https://www.facebook.com/artvinbirsevdadir/posts/1955801644671913/
65. Çoruh nehri üzerine kurulan büyük bir köprü (1935 July 3). Cumhuriyet Newspaper. Page: 1.
66. For more information on timber bridge: http://kopriyet.blogspot.com/2016/02/frat-ahsap-gecmek.html.
67. Nafia Vekaleti Railway Album No 000249 (1938) Ankara-Erzurum demiryolunun Sivas-Erzincan kısmının işletmeye açılma töreni hatıratından. Source: İBB Atatürk Library.
68. **Raşit Börekçi (1903–1977)** Graduated from Engineering School and worked in Nafia. He also served as a Parliament for two terms. He is son of Rıfat Börekçi who played key role in independence war and was the first president of Religious Affairs of the Republic of Turkey.
69. This bridge can be seen at: http://kopriyet.blogspot.com/2016/02/frat-ahsap-gecmek.html
70. Bridge name changed to Atatürk after surname law. The Surname Law of the Republic of Turkey was adopted on 21 June 1934. The law requires all citizens of Turkey to adopt the use of surnames. The nation's leader Mustafa Kemal was given the surname "Atatürk" (meaning: Father of the Türks) by the Grand National Assembly.
71. Alnar, G. (1939 May 28). Haliç'in Köprüleri: Gazi Köprüsü. Cumhuriyet Newspaper Archives.
72. Örmecioğlu (2010) Page: 158.
73. Alnar, G. (1939, May 28).
74. **Gaston Pigeaud** was a professor at the Ecole des Ponts et Chaussées and served as Inspecteur General des Ponts et Chaussées from 1928 until 1934. Pigeaud was one of the three members of the 1928 ad-hoc initiating committee in Vienna that led to the founding of

IABSE in 1929. He was also the inventor of the "Pigeaud" military truss bridge that was used in both World Wars.

75. Köprüyü Yapacak Mühendis Bugün Geliyor (1929 May 08) Akşam Newspaper. Page: 3.

76. Gazi Köprüsünün Planını Yapan Mühendis Mösyö Piju ve Emanet Heyeti Fenniye Müdürü (1929, May 14) Milliyet Newspaper. Page: 1.

77. Gazi Köprüsü İçin (1929, August 23) Vakit Newspaper. Page: 2.

78. Unkapanı Köprüsü İki Parça Edilecek (1930 January 6) Vakit Newspaper. Page: 3.

79. Figure 4.39 Orientalist elements on the first facade proposal of Pigeaud for Gazi Bridge, 1930. Source: Unclassified documents from the State National Archives-Republican Archives, KGM Fund, Binder no: 2095. Found in Örmecioğlu, H. T. (2010) Page: 162.

80. Gaip Aranıyor! Mösyö Piju'nun Projesi Nerelerde? (1930 February 28). Cumhuriyet Newspaper. Page: 1.

81. Emanet Köprü İçin Münakasa Açacak (1930 March 3) Cumhuriyet Newspaper. Page: 1.

82. Gazi Köprüsü Projesi (1930 April 2) Cumhuriyet Newspaper. Page: 1.

83. *Ali Galip (Galib) Alnar:* Chief Bridge Engineer at Municipality. He has published articles for Golden Horn bridges and was involved in the maintenance and repairs of these bridges. His known engineering project is the timber building, named as Yürüyen Köşk (The Moving House) in Yalova province, and the adjacent great plane tree. The house was moved 8m away on rails in order to protect the plane tree in 1929.

    Ref: Erbay, Ö.N. (2018). The Moving House or Atatürk House Museum and The Altered Mind-Set of Yalova Province. *Unimuseum*, 1:2, 28–33.

84. Gazi Köprüsü (1929, October 05) Akşam. Page: 4.

85. Gazi Köprüsü Projeleri (1937, March 20) Tan. Page: 2.

86. Örmecioğlu (2010) Pages: 158–167.

87. Örmecioğlu (2010) Page: 163.

88. Örmecioğlu (2010) Page: 163. Comment from M. Fuat the Director of Technical Services of the State Railways to the Ministry of Public Works on 10.12.1933.

89. Örmecioğlu (2010) Page: 164. Comment from Hilmi, the Minister of Public Works, to the Ministry of Internal Affairs, -.6.1933.

90. Örmecioğlu (2010) Page: 164. Muhiddin Üstündağ, the Mayor of İstanbul to the Ministry of Internal Affairs, 5.10.1933.

91. Örmecioğlu (2010) Page: 164. Correspondence between the Ministry of Public Works and the Ministry of Internal Affairs, 19.12.1933, source: Unclassified documents from the State National Archives-Republican Archives, KGM Fund, Binder no: 2094.

92. Örmecioğlu (2010) Page: 165. From the Ministry of Public Works to the Ministry of Internal Affairs on 14.12.1933.

93. Örmecioğlu (2010) Page: 165. From the Ministry of Public Works to the Prime Ministry on 23.12.1935

94. Alnar, G. (1938 December 11, 1938 May 13-15-17-25-28) Article Series: Haliç'in Köprüleri. Akşam Newspaper.

    Alnar, G. (1948, November 23) Column in the Newspaper. In this column, Alnar's letter to the newspaper was published. The letter is written to defence the condition of the bridge in a sentimental and interesting style.

95. Alnar's description are mostly driven from the preliminary design, which later revised by Pigeaud, therefore there are considerable differences. 1912 Bridge is a pontoon bridge in Galata location, also built by MAN. See Chapter 2 of this book for more description on 1912 dated bridge.

96. The balustrades of the bridge are made with the same motifs as the parapet and window motifs of the Ottoman Imperial New Mosque (Valide Sultan Mosque) situated at the western and of Galata Bridge.

97. Gazi Köprüsünün Mevkii (1930 June 30) Cumhuriyet Newspaper. Page: 1.

98. Gazi Köprüsünü Asma Yapmak Pekala Kabildir (1933 June 22) Cumhuriyet Newspaper. Page: 3.

99. Letter from J.A.L. Waddell on 31 August 1934. Source: Unclassified documents from the State National Archives-Republican Archives, KGM Fund, Binder no: 2094. Found in: Örmecioğlu (2010).

100. **Henri Prost,** who was one of the founders of town planning in France, was invited to prepare a master plan of İstanbul by the Turkish government in 1935.
101. Ayataç H. (2007) The International Diffusion of Planning Ideas: The Case of İstanbul. Turkey. *Journal of Planning History*, 6:2, 114–137.
102. Other three firms were: Fried Krupp A.G., Gutehoffnungshütte Werk Sterkrade, Dortmunder Union Brückenbau AG
103. **İbrahim Galip Fesçi** (1894–1956) graduated from civil engineering department of École d'Ingénieurs de l'Université de Lausanne in 1916. Fesçi had been selected as a member of Municipality Chamber. He has done contracting in Ankara, İstanbul and İzmit.
104. Atatürk Köprüsünün Temeli Atıldı (30 August 1936) Akşam Newspaper. Page: 1.
    Gazi Köprüsünün Temeli Dün Atıldı (30 August 1936) Son Posta Newspaper. Page: 1.
    Gazi Köprüsünün Temeli Atıldı (30 August 1936) Cumhuriyet Akşam Newspaper. Page: 1.
    Ataturk Köprüsünün Vaziesas Resmi (30 August 1936) Tan Newspaper. Page: 8.
105. First woman in İstanbul Municipality council and married to Mehmet Emin Koral, a high-ranking commander of the Turkish War of Independence. She was also the mother of the ceramic artist Füreya Koral.
106. Gazi Köprüsü İnşaatı (1936 November 20) Son Posta Newspaper. Page: 4.
107. Gazi Köprüsü İnşaatı İlerliyor (1937 March 10) Cumhuriyet Newspaper. Page: 4.
108. The number of piles used for the bridge is given as 490 by Alnar for both abutments and the same number was repeated by İlter. This newspaper provides pile numbers as "one hundred eighty and one hundred" respectively.
    From the sketches the number of piles appears to be around one hundred per abutment.
    The total number may have been mistyped in Alnar's article and should probably be 290 instead of 490.
109. Gazi Köprüsü Dubalarının İnşaatı (1938 March 12) Cumhuriyet Newspaper. Page: 1.
110. Number of pontoons are corrected to prevent confusion.
111. Yekta, R.Ö. (1939 May 23) Atatürk Köprüsü İnşaatı. Vakit Newspaper. Page: 7.
112. Different dimensions are given by Alnar and İlter. All dimensions given here are read from Arnold (1939) and (1940) unless noted otherwise.
    Arnold, P. (1939) Die Gazi Brücke über das Goldene Horn in Istanbul Der Bauingenieur v.20. Heft:15/16, 21 April 1939 Page: 204.
    Arnold, P. (1940) Die Gazi Brücke über das Goldene Horn in Istanbul Der Bauingenieur v.21. Heft:43/44, 20 November 1940 Page: 330.
113. In 1989, the bridge width was reconfigured, widening the carriageway from 16 to 20 m by reducing the width of the sidewalks to 2.5 m each.
114. Güngör, S. (1940 January 27) Milyonlara Mal Olan Zavallı Cisr-i Cedid. Cumhuriyet Newspaper. Page: 2
115. Cisri Cedid: meaning 'Old bridge' usually given name to Unkapanı bridge, which had been older than Galata Bridges since 1875. He uses the same expression for the new Gazi bridge to emphasize its bad condition just after its construction.
116. Güngör, S. (1939 November 01) Atatürk Köprüsüne Bakarken. Cumhuriyet Newspaper. Page: 2
117. BCA 29.12.1942 dated letter, 30-10-0-0_155_94_16_2- Ceyhan köprüsünün ayak kısmı ile demir üst yapısına ait sözleşmenin süresinin uzatılması
118. **Halit Köprücü** was a prominent bridge engineer and one of the founders of SAFERHA, which constructed the most bridges in both number and significance in Turkey. He later worked in KGM. See Chapter 3 for expanded biography.
119. BCA 10.02.1939 dated letter, 30-10-0-0_155_92_3_8-Ceyhan nehri üzerindeki köprü inşaatı sözleşmesine fiyat farkı eklenmesi
120. BCA 05.10.1938 dated letter, 30-18-1-2_84_88_20_2- Ceyhan ilinde kurulmakta olan Ceyhan köprüsü inşaatında kullanılacak montaj aletlerinin geçici kabul usulüyle yurda sokulması
121. **Davut Parker (Pistiryakof)** was a Russian immigrant. While working as an engineer in the service of Tsar Nicholas, he had to migrate to İstanbul due to the October revolution. After working as an engineer and subcontractor for a while, he started his own contracting business. His name Davut given to him by Ataturk, instead of David, and he changed his

surname with a court decision. His office was next to Akkaya's office in SAFERHA. Further information: Pertek Bridge ACH-27.

122. Örmecioğlu (2010) Page: 147.
123. Örmecioğlu (2010) Page: 149.
124. Cabinet decision for inviting offers from producer companies for construction of Mameki bridge 17.7.1946/4493. Source: Unclassified documents from the State National Archives Republican Archives', KGM Fund, Binder no:2315. Found in: Örmecioğlu (2010) Page: 148.
125. Örmecioğlu (2010) Page: 148.
126. https://www.bingolonline.com/haber/tarihi-genc-koprusu-yikiyor-48378.html.
127. *Sezai Türkeş* (1908–1998) a prominent bridge engineer of the early Republican era. Together with **Feyzi Akkaya** founded STFA Company, which was a leading engineering company in Turkey and worldwide. Their inventions and novel applications have been a vital contribution to the engineering field in Turkey. See Chapter 3 for expanded biography.
128. http://www.bartinhalkgazetesi.com.tr/bir-benzeri-daha-yok-4474m.htm.

Chapter 7

Beam Bridges and Unique Bridges

7.1 INTRODUCTION

The bridge collection to be presented in this chapter is beam and unique bridges built
in the early Republican era. Beam refers to the type in which the main structural element of the bridge is a beam. A definition of the beam is well provided by Trayona[1]
as: "... a straight, linear, horizontal or quasi-horizontal member supported on two or
more points. It supports the loads acting on it through its capacity to resist bending
or, ...".

Even though beam bridges seem to be the simplest and ordinary in appearance
compared to other types – arches, frames, trusses and cabled bridges – they prove
their diversity with many different options regarding their cross sections, span lengths,
materials, structural details and articulations. Their wide range of applications causes
beam bridges to have many sub-classifications. In the case of concrete beam bridges,
for example, small spans are usually crossed with solid slab sections, then as the span
gets longer beam systems are adopted. For even longer distances, pretensioned spans
using a box section become the solution.

Beam bridges are the most common type in modern bridge building since their
application is simpler and easier than other types. The first concrete bridges, however, were of arch type. The arch design had progressively evolved based on empirical
formulations and by trial-and-error methods, whereas the development of beam-type
needed theory to be employed for its structural design with reinforcement resisting the
tension.

The beam bridges described in this chapter are all made up of reinforced concrete. The unique structures covered in this chapter are also mostly of beam type.
The four exceptions to this are the Garzan arch bridge which is the first bridge of
the early Republic, Basmane bowstring bridge in İstanbul and Müstecap and Sivelan
frame bridges.

The bridges in this chapter are further grouped according to the designs or func-
tions of the structures. The three groups are namely, simple beam, cantilever (Gerber)
and landmark bridges. The grouping was made taking into consideration the most
distinct feature of the structure.

As expected, beam bridges dominate the bridge collection in numbers; for the first
10 years of the Republic, 20 bridges out of 48 were made of beam type. Regarding the
extensive list covering structures built until 1960, 150 out of 250 bridges were beam
bridges. These numbers represent 42% and 60% of their total numbers, respectively.

DOI: 10.1201/9781003175278-7

The reference list excludes most of the bridges built by Province and local authorities (municipalities). The above-mentioned percentages would increase dramatically if these bridges were also counted since the small structures built by provinces and local authorities were predominantly of beam type.

Another feature of beam bridges is their straightforward details and suitability for standardization. Therefore, they were convenient to be used as standard designs for bridges. The outcome of using standard designs led to the bridges being similar to each other, thus often not representing a unique character. Therefore, only the bridges with significant features and some representatives of the group are included in this chapter.

7.2   SIGNIFICANT FEATURES OF BEAM BRIDGES

The most practical and common beam bridge option is the "simple span", supported at both ends. In this type, the construction is relatively straightforward and the design is ready as it can be carried out by simple calculations. However, their inconvenience is in the joints along the travel surface causing a level of discomfort for users and causing them to be also prone to structural deterioration, therefore requiring regular maintenance.

Alternatively, the superstructure can also be built as a "continuous superstructure", which is jointless along the bridge deck, crossing over intermediate piers supporting the structure. This continuity may be applied all along or to certain sections of the bridge structure.

A further option arises with the modification of joints between superstructure and substructure to be made monolithic. This continuity leads to further adjustments for connections, for example, introducing haunches to the soffit of the beam to increase the cross-sectional depth and width of the joint to better spread the load and increase the capacity of the section.

Another group is the cantilever beam bridges. They are derived with the application of the cantilever construction technique for the beam-type bridges. In this method, the cantilevered spans project over the piers, then a suspended span rests on these cantilever ends. This cantilever beam bridge system is also named "Gerber Bridge" by the inventor, Engineer Heinrich Gerber, who patented this system in 1866 (Figure 7.1).

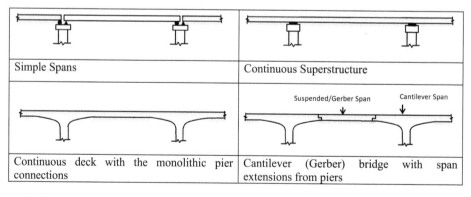

Figure 7.1 Beam bridge types (author).

Beam bridges are also classified according to the cross-section of the superstructure. The main types are slab, beam, girder and box sections.

Slab sections are the ones in which the superstructure is the solid slab, usually cast-in-place and reinforced to resist the loading. Generally, the slab section has solid concrete with a constant depth. In situations where the cross-sectional depths exceed certain limits, voids are formed to reduce the amount and self-weight of the concrete. The application range for this type can be regarded as small bridges up to a certain length which might usually be 10–15 m or up to 20 m depending on the overall structural design of the superstructure. For example, the cross-sectional depth decreases or longer spans can be achieved when the slabs are designed as a continuous system.

The slab section is an adaption from common building construction and therefore has a broad application. The main difference would be that the reinforcement in a building slab is supported on all edges, whereas a bridge slab has only two edges supported by piers or abutments (Figure 7.2).

Beam section is basically made with beams placed side by side at a predefined spacing and a slab on top forming a travel surface. The combination of beam and slab is called composite action, which refers to the cross section made up of a beam combined with the slab acting together structurally. The beams may have cross-sectional shapes such as **T**, **I** and **U** in common for concrete spans and are also used to name the sections.

The vertical element of the beam is the web and the remaining horizontal parts are called flanges.

Beam sections are a very basic component for superstructures and their variations are derived from applications like reinforcing or prestressing for concrete and construction methods like being cast-in-place or precast (formed elsewhere) and then placed in the structure.

Figure 7.2 **Single-span bridge built with slab–deck (author).**

T-beam appears to be the most common type and started to be in use as early as 1926 in Turkey. For the early Republican era, in the construction of bridges, nearly all beam sections were of T-beams with their upper flanges forming the slab of the bridge.

The construction of beam sections can be complicated especially for moulding the shape; however, its mass production and the nature of segmental construction, unlike the slab sections, made beam section favourable. They were constructed individually in a precast yard and transported to the site or precast on site. Beam sections also offered a considerable reduction in the self-weight of the structure, leading to longer spans.

Girder is the name used for beams with deeper cross sections. An alternative definition of girder is that girder is a beam, which supports the structure across its span and can also support other beams. On the other hand, the latter definition is also valid for beams. The distinctive description is not formulated between the beam and girder, and they can be often used interchangeably to refer to the same structure.

The span range for girders can extend a lot further, especially with prestressing techniques. Since all the bridges in this chapter are reinforced concrete, the girder section type is not observed for the period of the study.

Box section is the name for a deeper version of slab section, which forms voids named cells. An alternative definition of box section is the multiple beams used side by side with both their upper and bottom flanges interconnected. This section becomes exceptionally resistant to torsional effects because of its closed form.

In the early Republican period, box section is only observed in the Opera Bridge described as the last bridge in this chapter.

7.3   EARLY REPUBLICAN BEAM AND UNIQUE BRIDGES IN TURKEY

A total of 20 bridges are included in this chapter. They are grouped according to their structural types, features and outstanding characteristics. The four groups used are simple beam, cantilever (Gerber), landmark and unique bridges.

The simple beam group has three examples. The list starts with the Sarıoğlan Bridge, which is the first beam bridge design prepared by Public Works Authority/ Nafia.[2] The other two, Kırmastı and Çarşamba, were of significance being the longest beam bridge in the country at their time of construction. They both replaced previous timber crossings at their location.

The second group is dedicated to cantilever (Gerber) bridges, which are six in total, being the first Turkish applications of these structural types. All Gerber bridges built between 1923 and 1932 are included in this chapter. In this application, the approach of Nafia can be observed for the improvement and adaptation of designs for local conditions.

The next group of bridges, called landmark bridges, are the ones built as a result of urbanization in the cities and need to improve the networks and transport of the city with its surrounding. There are five bridges in total and are all built by local municipalities. Most of them are known to be designed with a strong architectural contribution. Three of these bridges are in İstanbul and the remaining two are located in the new capital, Ankara. The İstanbul-located bridges are Yoğurtçu, İstinye and Küçüksu; all demonstrate outstanding features and aesthetic appearance with their well-detailed parapets, bollards and facades. In Ankara are Etlik and Ziraat bridges, located on the main arterial of the city. These were built using the same design with different widths.

The remaining six bridges in this chapter are unique structures built in the early Republican era. Garzan was the first bridge built by Nafia; it is a stone structure with a single-span arch. Güzelhisar bridge was built with an innovative and experimental design approach, even though the outcome did not appear to be successful. This bridge is representative of the open-minded approach of engineers at that time and the supportive attitude of Nafia. The next bridge is Basmane Bridge in İstanbul. It is of bowstring type, with the distinctive feature of a triangulated truss arch member. The following, Müstecap and Sivelan Bridges are important representatives in terms of their structural types. The last entity in this chapter, Opera Bridge, is included in this study even though it is dated 1972, for its significance in Turkish engineering and architectural history with an outstanding design by prominent Italian Engineer, Luigi Nervi (Figure 7.3 and Table 7.1).

Simple Beam Bridges: The first simple beam bridge was Sarıoğlan, with a single span of 10 m, constructed by Nafia in 1931. Even though many more complex-type bridges had been built before, the construction of simple beam bridges only started in 1931. One of the reasons for this might be the efficient use of concrete material for large spans and the utilization of the more economical and already available materials like timber and stone for small spans.

At the same time, as observed in the Province Album,[3] smaller concrete bridges were already being built by local authorities. As stated in 1933 dated Nafia report[4]:

> the projects for small structures; like reinforced concrete culverts and bridges with length range from 60 cm to 14 m, were standardized and drawings prepared, so that the need for the provincial engineers to prepare these projects each time separately was eliminated… In this way, the construction of reinforced concrete bridges and culverts could be carried out by the provinces.

Therefore, this bridge design is presumably generated as a standard or typical design to be used by provinces.

Other bridges in this group are Kirmastı and Çarşamba Bridges, which were the longest bridges in Turkey at the date of construction.

BRD-01: Sarıoğlan bridge

This bridge is the first simple beam bridge constructed with a standard design prepared by Nafia. It has three spans, each of them 10 m, providing connectivity between Kayseri and Sivas crossing over the Sarıoğlan Creek. The width of the bridge is 6 m between parapets (Figure 7.4).

Sarıoğlan Bridge was needed since the former structure at this location collapsed with a flood on 12 June 1929. This emergency might have been the reason why the bridge was provided by Nafia and not the local province.

The tender was awarded to Engineers Hayri and Mustafa Efendi on 20 July 1930. The construction was completed on 14 March 1931.

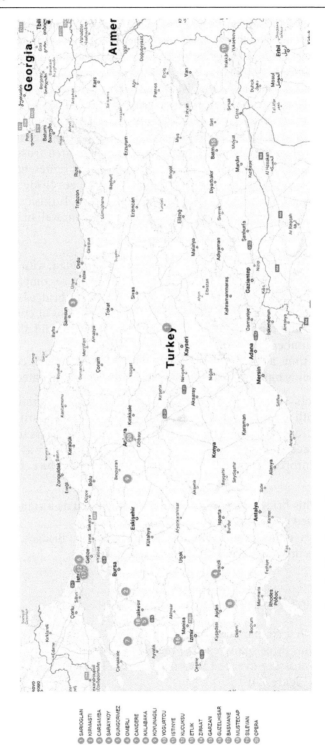

Figure 7.3 **Bridges shown in Google Map (author).**

Table 7.1 List of Early Republican Beam and Unique Bridges in Turkey (author)

BRD No	Name	Comp. Date	Location	River	No of Span	Main Span Length (m)	Status	Architect (a.), Designer (d.), Contractor (c.)	Significance	Coordinates
1	Sarioğlan	1931	Kayseri	Sarioğlan	3	11.2	Incomplete Information		First beam bridge design by Nafia	39.06757, 36.01574
2	Kirmasti	1927	Bursa Mustafakemalpaşa	Kirmasti	6	20	In use		Longest beam bridge when constructed in Turkey	40.03642, 28.41262
3	Çarşamba	1931	Samsun	Çarşamba	10	26	Standing	Reşit Bey	Longest bridge when constructed in Turkey	41.19833, 36.72426
4	Sarayköy	1928	Denizli	Menderes	3	30	Standing		First bridge in Gerber System	37.95262, 28.91887
5	Güngörmez	1930	Balikesir	Güngörmez	3	23.8	Standing	Saferha	Early application of Gerber System	39.61444, 27.54537
6	Ömerli	1931	Istanbul	Irva	3	20	Incomplete Information	Saferha	Early application of Gerber System	41.0871, 29.3336
7	Çandere	1931	Balikesir	Candere	6	20	Incomplete Information		Early application of Gerber System	39.99678, 26.98447
8	Kalabaka	1931	Aydin	Kalabaka	4	20	In use		Early application of Gerber System	37.616, 28.05729

(Continued)

Table 7.1 (Continued) List of Early Republican Beam and Unique Bridges in Turkey (author)

BRD No	Name	Comp. Date	Location	River	No of Span	Main Span Length (m)	Status	Architect (a.), Designer (d.), Contractor (c.)	Significance	Coordinates
9	Koyunağili	1932	Ankara	Sakarya	4	20	Incomplete Information		Early application of Gerber System	39.9874, 31.65358
10	Yoğurtcu	1925	Istanbul Kadikoy	Kurbağali	3		Incomplete information	a. Semih Rüstem	Architectural design	40.98321, 29.03424
11	Istinye	1928	Istanbul Üsküdar	Istinye	3		Standing		Architectural design	41.11372, 29.05426
12	Küçüksu	1928	Istanbul Kandilli	Göksu	2		Standing		Architectural design	41.07696, 29.06643
13	Etlik	1926	Ankara	Çubuk	3	12.2	Standing	a. Mimar Kemalettin c. Fahrettin Celal Nafiz Kortan	Architectural design	39.96029, 32.84723
14	Ziraat	1926	Ankara	Bentderesi	3	12.2	Demolished	a. Mimar Kemalettin c. Fahrettin Celal Nafiz Kortan	Architectural design	39.95108, 32.85836
15	Garzan Aviske İki Köprü	1924	Diyarbakir	Garzan	1	36	Replaced	c. Ali Raif Bilek	First bridge of Republic	37.96394, 41.34587
16	Güzelhisar	1927	IzmirAliaga	Güzelhisar	11	11	Incomplete Information		Unique pier design	38.83694, 26.99638
17	Basmane	1930	Istanbul Bakirkoy		1		Incomplete Information	a. Semih Rüstem	Unique design detail	40.98175, 28.88524
18	Müstecap	1931	Balikesir	Müstecap	3	12.8	In use	Hayri Bey	Unique bridge type	39.75214, 27.53243
19	Sivelan	1960	Hakkari	Zap	1	32	In use		Unique bridge type	37.70966, 44.05012
20	Opera Nervi	1972	Ankara	Atatürk Boulevard	4	45	In use	Luigi Nervi	Prominent engineer	39.93439, 32.85367

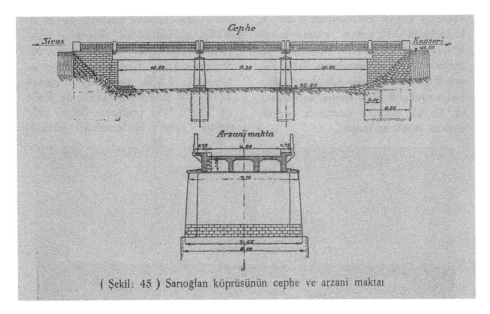

(Şekil: 45) Sarıoğlan köprüsünün cephe ve arzani maktaı

Figure 7.4 Sarıoğlan Bridge drawing from Nafia (1933).

BRD-02: Kirmastı (Mustafakemalpaşa) bridge

Kirmastı Bridge, also known as Mustafakemalpaşa Bridge from the name of the town, is located in Bursa Province. The 120-m long bridge is crossing over Kirmastı creek, with six spans of 20 m each. This bridge remained the longest beam bridge until the Çarşamba Bridge was constructed in 1931.

The construction of the bridge started in 1926 and was completed the next year, in 1927. Many previous timber bridges at this crossing were destroyed by floods.

This bridge has a continuous structural system with the superstructure constituted of T-beams and slabs. The superstructure is haunched at pier connections and the haunch is curved at the soffit to create a segmental arch profile. The piers are tapered from an enlarged base to the narrower width at their top. The top finish of the piers is detailed with circumferential capstones and a triangular fascia wall at the sides to enhance the appearance of the bridge (Figure 7.5).

The project was carried out by the local municipality; therefore, no information is found regarding this bridge in the Nafia reports or in the album[5] of province works. However, the bridge was recorded in the KGM list.

As learned from the municipality records,[6] the contractor completed the project without parapets, as it was not specified in the contract documents. Subsequently, a court case was concluded in favour of the municipality and the contractor was obliged to install metal railings. The name of the contractor (probably due to this conflict) is not mentioned in the documents (Figure 7.6).

Kirmastı bridge is still in use serving the local traffic.

Figure 7.5 Kirmastı Bridge seen in 1928 dated postcard (İBB Atatürk Library).

Figure 7.6 Kirmastı Bridge from KGM (1988).

BRD-03: Çarşamba bridge

This structure with its 274-m length became the longest concrete bridge in the country once it was completed in 1931, breaking the previous record 120 m of the Kirmastı Bridge. It crosses over the Yeşilırmak River in the town of Çarşamba, in the district of Samsun in the Black Sea Region of Turkey. This bridge connects a continuous network for the coastal road and provides a local connection within the town, which is located on both sides of the river.

Today, it only serves pedestrians and is one of the better-maintained bridges, embraced by its residents. External and additional triumphal arches were added at both

Figure 7.7 Çarşamba Bridge photo taken from the top of the municipality building (author).

entrances of the bridge in 2008. These arches are well detailed and add appeal to the bridge with their sympathetic design and compatible appearance.

The preliminary design for this bridge was planned to be made up of metal spans to carry both the Black Sea Coastal Road and Samsun–Çarşamba Railway on the same bridge; however, the coastal railway company couldn't meet its share. The idea was then abandoned and the design proceeded for a bridge to serve only road traffic. The bridge was tendered to İstanbul-based Engineer Reşid Bey's Company on 3 April 1928, and was completed in only 8 months (Figure 7.7).

The reinforced concrete bridge was designed in the cantilever system, with ten spans of 26 m each. Every second span has a suspended beam resting on cantilever extensions from piers. Confusingly, this bridge is not symmetrical in terms of span arrangements. All the Gerber spans have two joints; however, the first 26-m span on Terme (west) side starts with a Gerber span of one joint. The second span is a continuous beam, the following span has two joints and this order is sequentially repeated for the rest of the bridge. The last 26-m span is then with the continuous beam type.

The last spans on either end of the bridge have cantilevers of about 7-m long, adding another 14-m length to the structure. The total structural length of the bridge is 274 m. The bridge superstructure terminates with the beams resting on embankments.

The riverbanks at the ends of the bridge were elevated to provide road access on either side.

The width of the bridge was originally designed with a 4.80-m road width for vehicles and 115-cm sidewalks on both sides.

Initially, the pier foundation was planned to rest on a concrete piled system with a pile cap above the low water level. However, during construction, the driven lengths of the piles were observed to be insufficient against scour. Therefore, the four piers in the middle of the river were excavated further down with the assistance of sheet piles,

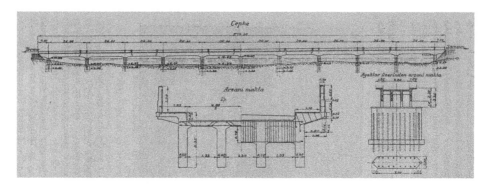

Figure 7.8 Çarşamba Bridge sketch from Nafia (1933).

and the foundations were placed on a lower and harder clay layer. The iron sheet piles were removed after the excavation and reused as a permanent protective curtain for the three piers, which were already constructed, on the western side. The piers on the eastern side were protected with wooden sheet piles. Additionally, all the foundations were protected against scouring by riprap.

The reinforcements for the superstructure were directly imported from Europe in 32-mm diameter bars to keep the number and consequently the connections as few as possible. The required joints were carried out with threaded screw sleeves (Figure 7.8).

The formwork was designed to be used in successive spans, thus making the formwork construction very economical. The deck was finished with a stone (parquet) pavement.

An ***innovative practice*** was implemented on this bridge and later repeated on other bridges of the early Republican era. Previously, a cantilever projection over the last pier was applied to lengthen the bridge and this cantilevered span was resting on conventional abutments. For example, Irva and Kirazlık arch bridges were designed with this feature. In this bridge, however, the last span made as a cantilever extension, did not rest structurally on the abutment, instead was self-supporting. Nafia described this situation simply by indicating that "the abutments are designed like piers". The system and the reason behind this new "solution" are not explained explicitly, other than some comments on its advantages like "'the reduction of mid-span moments at the adjacent span" mentioned in individual bridge descriptions. Nafia (1933)[7] further stated one of the reasons or outcomes behind this application in Güngörmez Bridge: "it was to avoid designing abutments with large dimensions to resist the soil pressure from embankment".

This new method was applied to many bridges and the difference between this "pier" and conventional abutment can be best observed in engineering sketches of the Koyunağılı Bridge. The Beypazarı (east) side of the bridge has a conventional bridge abutment and the other side on the Sarıköy (west) side is designed without abutment and the bridge deck continued as an embankment after the bridge.

The main reason behind this system would be to separate the bridge from an embankment. As also stated by Nafia when bridge descriptions are compiled together, the embankment does not carry "any" load from the bridge, meaning that the bridge structure is independent of the approaches.

In fact, in this application, the approach embankment level still needs to be increased up or retained at the bridge deck level, depending on the original road profile. In this regard, the system does not seem to offer much. However, the separation of the bridge structure and the embankment provides more design options and techniques to be employed for the embankment. For example, in the case of a conventional abutment, i.e., the bridge vertical loads transferred to the embankment, these loadings are carried down to a hard and competent stratum with an adequate capacity. Then, the approach fill needs to be excavated down to a reliable level or piled to reach this capable layer. Both cases will cause the abutment height to be taller. As the height of the abutment increases, the size of the structure increases exponentially.

On the contrary, if the embankment would not carry load from the bridge, the level of the embankment would be elevated by filling the existing ground layer with a compacted and selected soil, as long as stability and settlement requirements are fulfilled.

The overall efficiency of this engineering solution could still be discussed, as the two additional piers need to be constructed regardless. On the other hand, this application may certainly be considered a suitable solution for specific site conditions, especially when the stable embankment was already available at the design level of the road.

This application demonstrates how bridge building can be executed with expertise since taking risk and selecting an innovative approach over conventional methods require a proactive and responsive attitude from the authority to follow professionals to arise in proper environments.

The construction of this bridge was completed on 9 May 1931 (Figure 7.9).

Nowadays, the bridge appears to be modified in order to meet the requirements of pedestrians; ramps and stairs are added to the approaches for pedestrian circulation and the deck is paved with stone and levelled to a flat surface. The middle portion of

Figure 7.9 Çarşamba Bridge photo (Author).

the deck is allocated for cyclists and the outside areas are reserved for pedestrians. The drainage outlet appears to be shifted towards the centre from the edges of the bridge, presumably to prevent potential stains on the fascia.

The original parapet of the bridge was similar to other Gerber bridges – an ordinary metal railing continuous along the bridge with concrete posts at hinges of the cantilever span. Current parapets appear different consisting of metal railing panels, with concrete posts between each of them. Every second post also carries a lighting pole. Railings are decorated to match the style and surface ornamentation of the triumphal arches.

Another modification is the metal cladding added to the edges of the bridge covering the pipes, etc. along the bridge, and hiding nearly half of the depth of the superstructure as well. Clearing this cladding and pipes would notably improve the current appearance of the bridge.

Cantilever (Gerber) Bridges: Among the beam bridges, built by Nafia and KGM, nearly half of them are made as cantilever (Gerber)-type bridges. Gerber system offered vital advantages for Nafia, such as tolerating the differential settlements which could arise from ground conditions. Therefore, whenever the ground was softer than expected or there was not enough investigation made to determine the ground conditions beforehand, the cantilever type was chosen for the structural system. Another benefit of the Gerber type was the additional joints in the superstructure, making the structure more flexible and also meaning that the bridge is structurally determinate, which indicates that the design could be carried out with simple calculations.

The choice of the Gerber system was advocated by Nafia[8] as: "Almost all of the streams and rivers in Anatolia are full of residues and sediments. In such conditions, the Gerber system with reinforced concrete beams was very convenient...".

Between 1923 and 1933, six beam bridges were recorded being built with Gerber type with the first dated 1928. Some of these bridges shared designs; however, designs were improved or altered according to the local conditions.

The first bridge built with a Gerber system was the 1928 Sarayköy bridge, which was a three-span bridge with the Gerber beam in the middle span. The next was the Güngörmez Bridge constructed with the same design except with cantilever extensions of 7 m added to the end spans. The structures that followed were Ömerli, Kalabaka, Çandere and Koyunağılı in the order of date. These six bridges constructed during the first 10 years of Republic will be described individually in this chapter.

Six bridges were all constructed with the same type of cross section of T-beams. Generally, four beams were used to form the superstructure and the flange of the beam acted as a slab. The sidewalk slab is also structurally connected with the outer T-beam. All of the six bridges have similar parapet details with metal railing panels repeated with vertical steel posts in between. Concrete posts were placed at the joints. The identical concrete posts are also used at bridge ends and in some bridges at the piers.

In comparing the six Gerber-type bridges, the main difference between their details is the profile of the soffit of the beam in three different variations. Sarayköy and Kalabaka have a curved soffit in the middle span and the same curve extending into the Gerber beam. Güngörmez and Ömerli Bridges have a very gentle slope at piers and nearly a flat soffit within the superstructure. The suspended beam has a shallower depth than the bridge superstructure, creating a step in the profile. The other type in terms of soffit appearance is the continuous flat soffit along the bridge and very slight haunches introduced at piers as used in Çandere and Koyunağılı Bridges.

Nafia (1933) described the Güngörmez Bridge as a modified structure with additional improvements to the design, based on the earlier Sarayköy bridge. These improvements were mainly the revised drainage details to decrease the durability issues resulting from additional joints and better details for bearings.

After 1950, KGM standardized the types of beam bridges, including the Gerber types. Standardized projects were prepared both to feed the production boom and also to meet the load requirements and increase lane and clearance specifications for road infrastructure (Figure 7.10).

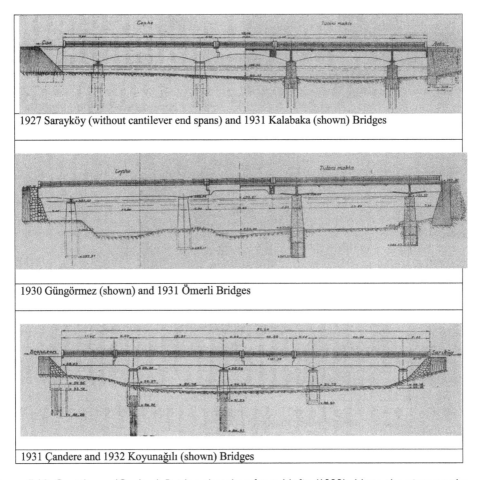

1927 Sarayköy (without cantilever end spans) and 1931 Kalabaka (shown) Bridges

1930 Güngörmez (shown) and 1931 Ömerli Bridges

1931 Çandere and 1932 Koyunağılı (shown) Bridges

Figure 7.10 Cantilever (Gerber) Bridge sketches from Nafia (1933). Note the view on the left side and the section on the right-hand side shown for the top two and the view is shown for both sides at the very bottom.

BRD-04: Sarayköy bridge

This was the first Turkish bridge designed with a Gerber system. It connected Denizli to Buldan, crossing over the Menderes River. Denizli has been an important centre on the trade route[9]: "The Denizli district was the most vibrant cloth production centre in western Anatolia during the later 19th century…The fame of the region rested on the output of two large nearby villages Kadıköy and Buldan…"

The bridge photograph can be found in Province Album.[10]

Nafia explained the reason for the preference for cantilever-type system[11]:

> … this bridge was built with a very useful system convenient for plain terrain and poor ground conditions. The bridge was built with the Gerber system and the suspended beam in the middle is 20 m long. Movable bearings were made of cast-iron and fixed bearings made of lead.

Spans are 25 m for side spans and 30 m for the middle span. The road width is 4.8 m between the kerbs with 60-cm sidewalks on either side, making the total width of the crossing 6 m.

The foundations were built using concrete caissons. The ground failed to achieve the required capacity; therefore, reinforced concrete piles were driven further down to reach a more suitable and competent foundation.

The bridge was tendered on 1 June 1927, and its acceptance by the authority as a new asset was executed on 26 December 1928.

BRD-05: Güngörmez bridge

Güngörmez Bridge is on the Balıkesir to Edremit Road crossing over the Güngörmez Creek with three main spans of 23.8 m each. The 13.4-m long Gerber beam is placed in the middle span.

As explained by Nafia[12]: "Since there was not a reliable investigation for the ground capacity, the superstructure designed in Gerber system to prevent damage of the bridge in case of differential settlements".

The innovative cantilevered span, as described earlier, was adopted in this bridge as an extension of the beams over the last pier. As a result, the bridge has three internal spans and another two 7m long spans at both ends. Therefore, there are four piers in total, instead of two piers for a three-span bridge.

The width of the road surface is 4.8 m with 60-cm sidewalks on both sides resulting in a whole width of the bridge as 6 m.

Movable bearings were made of cast iron and fixed bearings were made of lead. In addition, two layers of 2-cm thick lead plates were placed under the Gerber beam as joints.

In order to prevent the rainwater from draining through the joints, a 1-cm thick asphalt layer was laid on the deck together with a 4-cm screed gravel layer as protection. The bridge profile was made flat; therefore, a crossfall, i.e. the sloping of a roadway towards the shoulder on either side, was provided in the transverse direction to ensure drainage of rainwater.

The bridge was awarded to SAFER company with a contract dated 10 October 1929, and completed on 20 December 1930.

BRD-06: Ömerli bridge

Ömerli Bridge is located in İstanbul Province on the Üsküdar to Şile Road crossing over the Irva Creek. It replaced the previous timber bridge at the same location (Figure 7.11).

The superstructure details are the same as the Güngörmez Bridge except for the span lengths. The three main spans are 20 m, and there are 6-m long cantilever extension spans at both ends, totalling 72 m. The suspended span in the middle is 11.2 m long.

It was contracted on 3 December 1929, and completed on 22 August 1931.

BRD-07: Çandere bridge

This bridge is on the Balya to Çanakkale Road, crossing over the Çandere Creek, which is known to frequently discharge a significant amount of water. Therefore, it was deemed necessary to construct a permanent reinforced concrete bridge extending over the flood plain. The structure has six spans in total, with 19.4-m long end spans and the remaining 20-m long four spans over the stream. This bridge has three Gerber spans positioned at the first, third and fifth spans starting from the Balıkesir (east) side to Çanakkale (west) side. The suspended beam lengths are 16.4 m for the first span and 12 m for the remaining Gerber spans. The bridge width between the parapets is 6 m (Figure 7.12).

The tender was awarded on 29 July 1930, and the construction was completed on 30 August 1931.

BRD-08: Kalabaka bridge

This bridge on the Aydın–Muğla Road has similar details to the Sarayköy bridge (Figure 7.13).

Figure 7.11 Ömerli Bridge photograph from Nafia (1933).

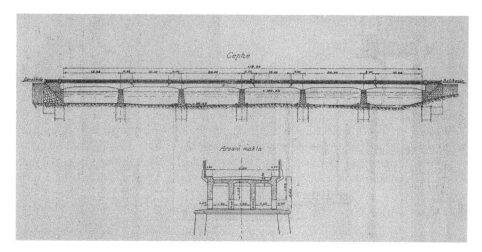

Figure 7.12 Çandere Bridge sketch from Nafia (1933).

Figure 7.13 Kalabaka Bridge photograph from Nafia (1933).

It has five spans in total with 20-m long main side spans and 18 m in the middle. The end span cantilever extensions were of 7 m. The suspended beam in the mid-span is 11.2 m long.

This structure was completed on 10 August 1931.

BRD-09: Koyunağılı bridge

Koyunağılı Bridge is crossing over the Sakarya River on the Ankara-Beypazarı-Sarıköy Road.

This bridge has five spans each with different lengths. Gerber spans are the first and third spans starting from Beypazarı (east) with 11.45 and 11-m suspended beams, respectively.

The construction was contracted on 2 September 1930, and was completed on 13 February 1932.

Landmark Bridges: Modernization and rebuilding were the overall drive in the whole country from the local city to the global country. The municipalities were pre-paring urbanization plans, providing a clean environment, organizing the urban areas, and designing the city to create a space for modern and egalitarian life for all. The cities, especially Ankara, İstanbul and İzmir, had master plans prepared to reorganize their districts with urbanized centres and modern living spaces.

Therefore, it is not surprising to have some noteworthy bridges appearing in these cities. They are presumably planned for an aesthetically pleasing appearance with at least some standard architectural input, depending on the nature of the project and the location of the bridge. Even though the original intention might not be to create an iconic structure, these bridges certainly had a positive impact on residents and their vicinity to the extent of sometimes being used as scenery in photos. Nowadays, they are a representation of their time and significance with their simple structure and interesting details, even though many of them do not exist anymore.

Although the landmark bridges were very probably more in numbers and very diverse in types, only some of them were found by the author. These are namely Küçüksu, İstinye and Yoğurtçu beam bridges and Basmane bowstring bridges de-tailed, all in İstanbul, as well as Etlik and Ziraat Bridges on the main arterials of the capital Ankara.

The architect of some bridges in İstanbul was Semih Rüstem as found in a KGM document.[13] In this document, which refers to the period of 1930–1933, a contrac-tor is applying for the bridge tender over the Gediz River and listed the company experience through the references carried out by Architect Semih Rüstem. The bridges listed were Basmane Bridge in Bakırköy, Yoğurtçu Bridge in Kadıköy, Kağıthane (Sünnet) and Silahtarağa Bowstring Bridges and Aslan Bridge on the Ereğli-Devrek Road in Zonguldak Province.

Another letter attached within this KGM document dated 12 August 1930, lists Basmane and Yoğurtçu Bridges as completed by the contractor. The name of the contractor cannot be read from the documents.

Even though İstinye and Küçüksu Bridges were similar to the Yoğurtçu Bridge, these two are not listed in the above-mentioned document.

The available information about these bridges is very limited; however, they will be described here so that they can be recorded for future studies.

BRD-10: Yoğurtçu bridge

Yoğurtçu Bridge was in Kadıköy, İstanbul, next to the park with the same name. The Yoğurtçu Park was constructed by the municipality as a part of landscaping work in the vicinity, which was initiated by donations of Süreyya İlmen[14] and finished in 1925.

The beam bridge had three variable spans with approximately 5-m outer spans and 20 m in the middle, as seen from photographs. The beam soffit is curved, resem-bling a segmental arch, and became the centre of attention for the viewers, making the structure appear higher than if the soffit had been made straight. The ring of the arch was grooved in imitation of the voussoirs of a stone bridge (Figure 7.14).

This bridge had a heavy balustrade with concrete posts placed between the parapet panels. The parapets had one panel at the outer spans and five panels used for the middle span. At the ends and at piers, a recessed panel was placed accentuating the height of the bridge. The recessed panels are higher than the parapets.

The same panels and posts were also repeated at the stairs on the eastern side. The stairs landed on the riverbank, where a toilet was also provided. Metal poles were mounted on top of the entrance posts for lighting purposes. The poles were adorned with iron embellishments on top.

The bridge had a white marble plate indicating its name and date, placed at the crown in the middle span below the balustrade.

Only photographs are available for this bridge. No documentation could be found and the structure was also not found in its probable location.

BRD-11: İstinye bridge

This three-span bridge carries İstinye Street through the İstinye neighbourhood of Sarıyer District of İstanbul. The bridge was built in 1928 from the date inscribed on the bollard at the entrance.

The architectural features of the bridge are significant with parapet panels and bollards giving character to the bridge. The parapet of the bridge has repetitive architectural features. There is one parapet panel per span and another recessed panel is placed at the pier locations, on either side of the deck.

The four bollards placed on both sides of the entrances extend 2.4 m above the bridge deck. They are column-like square bollards and their exterior was shaped to imitate the appearance of masonry courses. The finish of the bollards is made with three layered capstones and a small sphere was placed at the tip. These bollards have also niches on their exterior sides. The parapets turn around the bollards at the entrances of the bridge (Figure 7.15).

One of these bollards has an identification plate with the name and date of the bridge carved on it.

The bridge structure is a continuous deck with two slim piers, which are also decorated with an imprint to look like a masonry structure.

Additional spans are located at both approaches, before the entrance bollards. These spans are small culvert openings. The facade for the bridge again is moulded to imitate the appearance of masonry courses on the spandrel walls.

The structural beam does not have any treatment to its surface finish in contrast to the rest of the masonry imitating sections.

The appearance of the bridge is ordered, basic and clear with a flat soffit and seemingly equal-length spans. The continuously adorned concrete parapets along the bridge create a bold straight line, and high bollards standing at the entrances emphasize the height. Therefore, the bridge represents harmony between horizontal and vertical elements and simple and pronounced surface textures; all in accordance with each other. Thus, the bridge attracts one's eye with its unity.

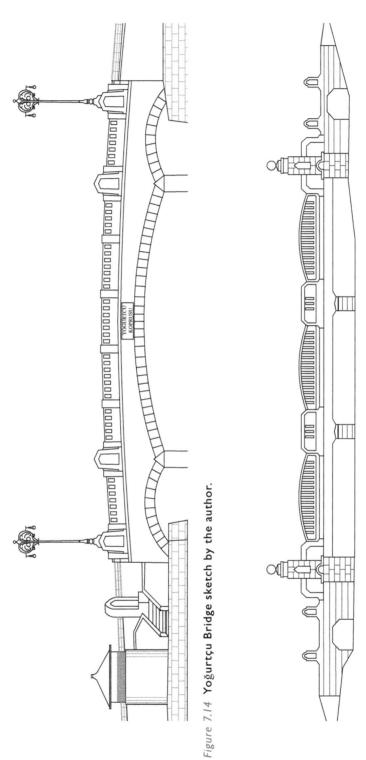

Figure 7.14 Yoğurtçu Bridge sketch by the author.

Figure 7.15 İstinye Bridge sketch by the author.

The current condition of the bridge is not known. All the information explained here is derived from images.

BRD-12: Küçüksu bridge

The bridge is on the Kandıra Road in the Göksu neighbourhood of Beykoz District in İstanbul. The construction date is known as 1928 from its inscription plate, which is fixed on the parapet with the name and date written in black paint on white marble.

There is also an album[15] that collected photographs of the bridge during its construction and opening event. This album was prepared in Ottoman and French languages and was presented to Kandilli Mayor (Figure 7.16).

The bridge is simple and elegant in its original appearance. There are two spans resting on a slender pier in the middle. Indeed, the even number of total bridge spans is not generally favoured in aesthetic bridge design since the view is disturbed by the pier placed in the middle of the landscape. However, in this bridge, the overall design is successfully accomplished to provide an appealing appearance. The slim pier in the middle is balanced by the robust wing walls on either end of the bridge. Both the pier and wing walls are decorated in imitation of masonry structures, in contrast to the rest of the plain-surfaced bridge elements.

The superstructure appears to be made with individual beams. The soffit profile of the beam is flat with haunches introduced at the pier and abutment connections.

The parapets are solid and continuous along the bridge, with ornamental openings shaped like pointed arches. The parapet posts, nearly flush with the parapet wall, are placed in the middle of the bridge above the piers and just adjacent to the abutments. The entrances are accentuated with wider and taller panels, which are placed

Figure 7.16 Küçüksu Bridge during opening (SALT research, photograph and postcard collection).

on the top of the abutment wing wall. Similar but smaller panels are repeated at the approaches of the bridge.

The bridge still stands with partly collapsed parapets on both sides. It is still being used for local vehicle traffic. The original parapets on the east side are protected with metal railings introduced on the carriageway side. On the west side, parapets are also still standing and exposed to direct traffic.

BRD-13 and BRD-14: Etlik and Ziraat Mektebi bridges

Etlik Bridge was built across the Çubuk Stream on the road to Etlik and Keçiören Vineyards in Ankara. Ziraat Mektebi, meaning agriculture school, was named after the school nearby and was on the road from the city centre to Dışkapı.

Etlik and Ziraat Bridges are the only examples of neoclassical architectural applications of bridges in Turkey. These structures are the sole representative of their time and style, designed by a prominent architect named Kemalettin.

Both bridges were identical with the exception of the width. Etlik Bridge is 8 m in total width, 6 m of clear roadway and 1-m sidewalk on either side. Ziraat Bridge had a total width of 12.2 m with a 10-m clear road between sidewalks (Figure 7.17).

Today, Etlik Bridge is still in use while Ziraat Mektebi Bridge was demolished during a municipal operation in the 1950s. Therefore, Etlik Bridge is focussed on here.

Etlik Bridge has a distinctively ornamented parapet design and an unusual arched profile for its spans resembling a basket-handle arch type. The design is vibrant with rich details over a relatively small area of the bridge.

The bridge is straight with three continuous spans. The outer spans are 10.50 m with 15 m for the middle span. The superstructure beam has been curved as a five-centred arch. The soffit is segmented by five elements symmetrically arranged to form an arch shape. The arch starts with a short segment, followed by an adjacent transition curve to the middle portion of the flat alignment. This flat middle segment is nearly half the length of the span. This arch is similar in appearance to a basket-handled arch, which is a three-centred arch having a crown with a longer radius. The arch is purely an architectural feature whilst the main structural element is a reinforced concrete beam. A similar arch pattern is repeated underneath the parapets along the bridge.

The parapets follow a pattern along the entire bridge. The posts at pier locations and abutments are higher and finished with a capital. The sidewalk and parapets form a projecting platform at the sides of the bridge with an ornamented profile. This profile commences about a metre above the cutwater of the pier for the posts at piers and abutments.

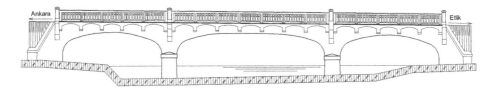

Figure 7.17 Etlik Bridge sketch by the author.

As seen from the KGM design archive drawings,[16] there are some minor differences between the design drawings and the built structure on site. For example, in the original drawings, the posts at pier locations are topped by an architectural finial; however, this detail is omitted in the built structure.

Structurally, the bridge is an integral design with superstructure and substructure as a continuous system. Nafia (1933) further explains: "Vertical reinforcement was placed in piers to prevent the piers from cracking under the influence of horizontal forces" (Figure 7.18).

The architect of the bridges was Mimar Kemalettin,[17] who was known for his design of public buildings in Ankara and İstanbul. Mimar Kemalettin (Kemalettin the Architect, 1870–1927) was a renowned architect of the very-late Ottoman period and the early years of the newly established republic. He was among the pioneers of the first national architectural movement. This movement, inspired by Ottomanism, sought to capture classical elements of Ottoman and Seljuk architecture and use them in the construction of modern architecture.

The construction of the bridges started in 1925 and was completed a year later. From KGM documents, the name of the contractor was Fahrettin Celal. However, Örmecioğlu (2010) provides information regarding the contractor as: "Construction began in 1925, most probably under contractors of Erzurumlu Hacı Ahmedzade Nafız (Kotan) Bey, and the bridges were opened to traffic in 1926".

Ziraat Bridge was demolished as part of the urban planning of the city of Ankara in 1955.[18]

Etlik Bridge was deteriorated and exhibited poor condition with no parapets, but was restored to its original condition in 2018 by KGM, as observed during the author's visit in 2019.

Figure 7.18 Etlik Bridge after its restoration (author).

Unique Bridges: Bridges are usually considered unique if they are very attractive or have extraordinary features; however, in this book, the unique bridges are those different from others that did not fit in the other chapters of the book. This section includes five bridges in total.

Garzan is the first bridge built by Nafia; unfortunately, it does not survive today. Güzelhisar, Müstecap and Sivelan bridges are the rare representatives of their types and therefore included separately in this section. Sivelan dated 1960 and Opera Bridge dated 1977 were built by KGM and the municipality, respectively. These two structures are not strictly early republican bridges; however, given their heritage significance, they have been included in this book.

The current condition of Güzelhisar is not known. The remaining three bridges Müstecap, Sivelan and Opera are still standing and all servicing vehicle traffic.

The final addition to unique bridges is the Basmane Bridge, a bowstring bridge with an unusual arch form.

BRD-15: Garzan (Aviske, İkiköprü) bridge

Garzan Bridge was the first bridge of the Republic. Unfortunately, the bridge is no longer standing but its history tells us about the conditions under which it was built. The bridge was on Awiski Stream, today called Yanarsu, and was located on the Diyarbakır to Siirt Road in the province of Siirt.

It had a single span of 36 m. It was different and innovative with its segmental arch which appears to have had a quite low rise. Segmental arches are those having less height than a semicircular arch style.

The connecting road level required a very low rise for the arch, and a segmental type, which would exert more thrust on the abutments, was chosen. Considering the existence of solid rock for both sides of the bridge to be supported on, the segmental arch was certainly a reasonable decision (Figure 7.19).

The bridge was made up of stones interconnected with mortar; however, the cement required for the mortar was not easily available to the site because of the restricted access. Therefore, the masonry is made with "Khorasan mortar",[19] which is a combination of ground brick or natural stone and lime used to bind the content. This method has been used since the time of the Roman Empire and the binder used varied according to region.

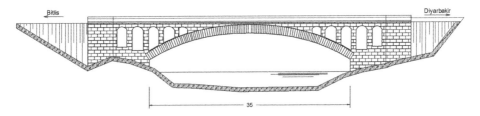

Figure 7.19 **Garzan Bridge, the first bridge of early Republican era (author).**

Bridge had five openings within each spandrel mainly to reduce the weight, use less material and also to provide more passage for allowing excess water during flooding. The bridge is simple and symmetrical.

The chief engineer of Nafia at that time, Kemal Hayırlıoğlu,[20] explains the difficulties during bridge construction in the following lines:

> This first example led to disruptions on the bridge due to inexperience and technical deficiencies. The scaffolding posts were placed on a lime mortar masonry pier which was placed directly on the sand in the stream bed. The construction had continued until the winter season and after the construction was finished, the mortar had to be supported for a longer time to harden. Therefore, the scaffolding was left in place, however, a sudden flood caused the scaffolding to collapse due to scour of the post in the river. Due to this failure, the arch was also damaged. During the repair of the bridge, a new scaffolding was constructed and the arch was completely demolished and reconstructed. In order not to damage the bridge again due to floods, the pier was lowered 54 days after the arch was closed and there was no problem. However, since the mortar was not fully hardened, a deflection of 125 mm was measured in the crown of the arch. Although this bridge, which was built under very primitive working conditions, suffered such damage, it was a start for bridge building activities in the early Republican Period as it achieved a large span of 36 meters.[21]

A photograph of the bridge shows the construction stage and a footnote on the photograph states[22]: "the joints filled in" with the date also noted as 24 November 1923. Considering the information provided by Hayırlıoğlu and the construction timeline of the bridge, the photograph must have been of the structure which collapsed in the first attempt (Figure 7.20).

The resident engineer of this bridge was Ali Raif Bilek, who was described by Feyzi Akkaya[23] in his book[24]:

> Raif Bey was the old one with his beak nose. They called him the Coachman Raif in school. He was a man of principles with hard and inverse manners, from whom most stayed away but I warmed to him over time... In the rebellion of the Sheikh Said, he built the masonry Aviski Bridge in Garzan Stream. There, he was respected like a Kurdish prince. While the Money is not valid other than gold and silver, even the Sheikh Said's side did not deal with Raif Bey's mule convoy; which announced: "mail to engineer". The little, ornate tomb of his daughter who died there is at the head of the bridge.

The construction of the bridge started in 1923 and was completed in the following year. The photograph of the bridge is given in a newspaper dated 1946. It has since been replaced with another bridge (Figure 7.21).

Figure 7.20 Photo with a handwritten note: "Garzan Bridge, the joints are filled in on 14 November 1923" (İBB Atatürk Library).

Figure 7.21 Garzan Bridge with the railway bridge in the background (25 July 1946 dated Ulus Newspaper).

BRD-16: Güzelhisar bridge

Güzelhisar Bridge was crossing Kocaçay, today called Güzelhisar Creek, on the 36th km of the İzmir–Bergama Road. The significance of this bridge relies on the technical features of the substructure design. As explained in the Nafia report, the initial plan was to standardize the design of the bridge; therefore, it is not surprising that it was not applied again.

Nafia explains the concept as:

> Many rivers channel and creeks in Turkey, dry out or have very little water for most of the time except when the heavy rains occur and floods which happens in few days of the year. At the same time, a considerable number of bridges are required to ensure transportation in all seasons. In order to provide a feasible and practical solution, it is proposed to apply an inexpensive and easy-to-construct structure that complies with the regime of the streams. In this regard, Güzelhisar Bridge is an example of an economical design, which meets the requirements, specifically for the wide stream with no debris or large branches carried during an unexpected flood (Figure 7.22).

This bridge has 11 spans, each of them 11 m long. The total length of the structure is 121 m, and the width is 6 m between railings. The cross section of the bridge differs from the usual with only two edge beams carrying the whole deck. The flange of the edge beam is also part of the slab and sidewalk of the bridge; all the elements were detailed structurally to work together.

The pier consists of a total of six driven piles each, symmetrically and equally spaced on the up and downstream sides of the bridge. The piles form two triangles on the plan, with two piles located alongside the bridge and the third pile being driven further away from the deck.

Piles were finished and connected with a cap beam just below the deck level of the bridge. Therefore, pile cap and pier cap are the same in this structure. The main structural edge beams of the superstructure were supported on this pile cap.

Figure 7.22 Güzelhisar Bridge photograph from Nafia (1933).

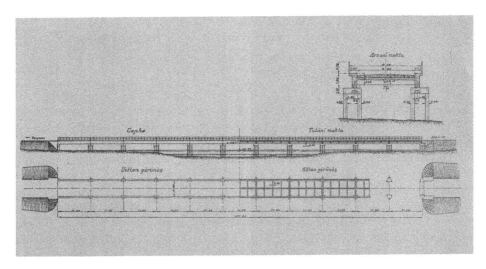

Figure 7.23 Güzelhisar Bridge sketch from Nafia (1933).

The design was efficient in the aspects of reducing the effects of scouring and accumulation of debris, having as little disturbance as possible to the water flow. A continuous superstructure could be adopted because the foundation system was stable with piled system preventing differential settlements. Abutments were also constructed with the same piling system as seen from the sketches of Nafia (Figure 7.23).

This bridge design exhibits an unusual approach and it is not clear where the concept was derived from. Later, when reviewing in parallel with the Yahşihan and Gediz Bridges,[25] where both used metal screw piles for their pier design, the intent in Güzelhisar Bridge becomes evident. This bridge seems to be a concrete application of the same design feature in terms of piers.

Güzelhisar Bridge was tendered in June 1926 and was completed in February 1927. Therefore, it took approximately 9 months to construct the bridge. The cost of building this structure reached a value of 57.450 lira, which could be considered economical compared to other similar bridges built at that time. The cost for this bridge was approximately 34% lower than the next lowest cost Adagide Bridge, if the cost per length of the bridges is compared.[26]

As learned from Nafia's (1933) report, the piles were driven to 6 m depth into the ground with difficulty. Another challenge would be the driving of the precast piles within allowable accuracy because the deviation in alignment would continue in the upper column section above the pile. At the same time, the structural and buckling height of the substructure would increase with this design compared to the design with pile cap option.

When the photographs of the bridge are examined in close detail, it can be noticed that the columns at the piers are rectangular in their cross section. On the contrary, the design sketches of Nafia show a circular cross section for both the pile and the column. Nafia might have initially intended to drive longer piles so that the upper portions would be utilized as a column. However, this presumably would have not been possible or practical for construction. Then, it is inferred that the upper portion of the piles may have been completed with cast-in-place concreting.

The current condition of this bridge is not known to the author.

BRD-17: Basmane bridge

This single-span bridge provided connectivity in the Bakırköy District of İstanbul. The name Basmane comes from a local textile factory, which was later renamed as Sümerbank.

The structure uses a bowstring arch type, as indicated in the letter found in the KGM archive. From this letter, it is understood that the architect of the bridge was presumably Semih Rüstem.

The tied-arch bridge type, also called bowstring arch, was first observed in Bursa as a single-span bridge dated in 1912 and is one of the early applications of this structure of road bridges. Bowstring bridges are covered in detail in Chapter 4 of this book.

Basmane bridge with its main structural arch formed as a triangulated truss is a rather exceptional case (Figure 7.24).

One of the rare known examples of a similar design is dated between 1925 and 1927 and is located in the Mellègue Valley in Tunisia. The Mellègue bowstring bridge was an engineering feat designed by Henry Lossier for the Fourré & Rhodes Company. This bridge reached a 92-m span, with about a 15-m rise as a tied arch, in which the arch was formed as a triangulated section (Figure 7.25).

Basmane bridge span was approximately 15 m long as seen from its photographs. Eight hangers were placed at equal intervals and the entrances were emphasized with well-detailed springing. Metal railings used between hangers were also detailed along the bridge. Two lateral beams connected the main arches. These lateral beams were pleasing with a straight member on top and a curved member at the bottom, connected

Figure 7.24 **Basmane Bridge photo (author's collection).**

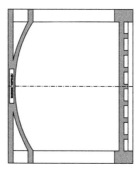

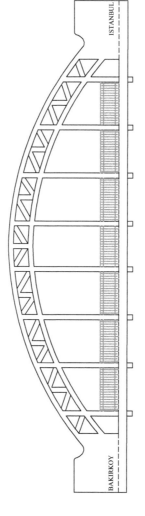

Figure 7.25 Basmane Bridge sketch by the author.

to the top and bottom flanges of the main arch, respectively. The name of the bridge was written on a white plate, placed on one of the lateral beams.

There is no additional information found by the author other than the photographs.

BRD-18: Müstecap bridge

Müstecap Bridge is on the Balıkesir–Çanakkale Road. The significant feature of this bridge is the use of its structural type of frame, as a rare application in the early Republican era and even in general bridge-building practice in Turkey.

During the construction of the bridge, hinges are formed on the bottom and top of the piers. The stability of the piers was provided by the cabled supports, also preventing the superstructure from load exposures during construction. Once completed, these hinge joints were converted into rigid connections at the end of the construction (Figure 7.26).

Since solid ground was identified at a shallow depth, it was possible to design the structural system of the bridge as a continuous beam.

The bridge is considered frame type and the structural system is described as a "pendulum" in Nafia's (1933) report and favoured as movable joints were prevented in the structure. However, the monolithic behaviour and load sharing between beam and piers are relative to their dimensions, with very slender wall-type piers; this locking-in effect may not be very effective in practice.

As the river carries large stones during a flood, the lower section of the pier within the river was enlarged to increase its area against the water pressure. Considering that floods could also tree branches, the piers were built as a solid wall in the transverse direction, without internal openings that could catch floating branches (Figure 7.27).

There are four beams topped with a slab forming the 6-m width deck including pedestrian sidewalks. The bridge spans are 12, 12.80 and 12 m with a symmetrical

Figure 7.26 Müstecap Bridge photograph from Nafia (1933).

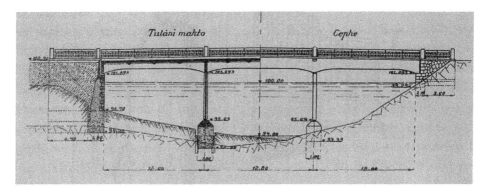

Figure 7.27 Müstecap Bridge sketch from Nafia (1933).

Figure 7.28 Sivelan Bridge photo (KGM Calendar).

arrangement. The excavation for the foundation reached down to a rock layer using diaphragm walls.

The contract dated 30 July 1930, together with the Agonya Bridges, was signed with Hayri Bey, and the construction was completed on 25 November 1931.

BRD-19: Sivelan bridge

This bridge was built on the Hakkari–Van–İran Road in 1960 to cross over the Zap River. It is one of the rare examples of a bridge built with a frame system.

The bridge is placed at the Yüksekova Junction, where there is also a police station with the same name next to it (Figure 7.28).

A frame bridge system with an inclined leg in this structure is an option between arch and beam bridge types in terms of the structural behaviour, mainly based on how the horizontal forces are restrained in the system. The frame type can cross longer spans than the beam type with the same superstructure section. The advantages over the arch type might be that the minimum headroom required under the bridge is less compared to arch design and also that the formwork system can be more practical for frame bridges.

Sivelan Bridge looks simple and elegant, especially with how the continuity of the structure and the leg details contributes to the pleasing appearance of the bridge. This bridge has four separate portal ribs arranged to form the bridge. The soffit of the structure has a curved profile, resembling an arch bridge.

The dimensions of the structural elements of the frame were determined in proportion to the load they would carry. The narrowing of the section of the leg towards its base indicates the existence of a hinge at this location. Conversely, at the top of the leg, where it joins with the deck of the bridge, the leg section increases, where also the flexural moment becomes larger. The leg shape and bridge overall capture the attention of the viewers with their variable dimensions.

The total length of the bridge is given as 32 m in KGM documents.

BRD-20: Opera (Nervi) bridge

Sydney, Australia, is not the only place where an Opera House comes with a bridge. Ankara Opera House also hosts a bridge, one designed by Luigi Nervi!

Both the bridge and the building have become precious contributions to the identity of the city. The Opera House was originally designed as an exhibition hall by Şevki Balmumcu in 1933 but was later transformed into an Opera House by Paul Bonatz (Figure 7.29).

Figure 7.29 Opera Bridge east abutment from north looking south (author).

The bridge might be regarded as a city sculpture rather than a bridge with its robust size, unique design and the distinction of being a "Nervi" structure. Luigi Nervi from Italy was an outstanding engineer who created amazing and unique structural designs for aircraft hangers, stadiums and fortunately some bridges, one of which is in Ankara.

This four-span bridge forms one of the key junctions of Turkey's capital, located over Atatürk Boulevard, the main arterial of the city. The bridge crosses over the boulevard and its collector roads on either side carrying the main arterial of Talatpaşa Street. This location is also very close to the old city centre, Ulus, where the first Republican parliament took place.

After a long search for the history of the bridge, information was found by Örmecioğlu[27] as given below:

> ...late 1965. In that year, they [City Council] decided to commission the project [Opera Junction] to a renowned foreign designer in order to get a significant urban landmark.[28] Haluk Alatan, the chief expert of the Ankara Metropolitan Planning Office, recommended Nervi, whom he had met while working in Luigi Piccinat's office during his doctoral studies. ... Antonio Nervi, the son of Pier Luigi and chief engineer at the Studio Nervi, visited Turkey in August 1968 and the office was officially commissioned by the end of that year...The Opera Bridge project was prepared in 1969; after minor changes, construction began in early 1970 and was completed in 1972. The Municipality took special care to realize the project in its original details and created a dedicated budget for it...

Another letter[29] found in the Republican archives states that Nervi was appointed to work for the Municipality of Ankara from 20 June 1969, to 20 June 1970.

There were no records or documents available for the Opera Bridge, so the observations made during the author's site visits in 2018 and 2019 are used here to describe the structure. Opera Bridge is similar to the 1968-dated Risorgimento[30] Bridge designed by Nervi. Risorgimento Bridge has three spans of 34.50, 62 and 34.50 m lengths, crossing over River Adige in Verona (Figure 7.30).

Figure 7.30 **Opera Bridge from a distance panoramic view (Photo: M.Arch. Sevil Işıl Ören).**

Opera Bridge has four spans with dimensions of approximately 25, 45, 45 and 25 m measured from images and maps. The superstructure is a multicellular concrete box. The change in the structural depth is not obvious for this bridge, whereas it is very recognisable in Risorgimento Bridge.

As noted, the Opera Bridge has a flat traffic level and soffit along the bridge; however, Risorgimento has a curved profile for the deck and soffit. The main variant from the exterior view is the slope of the sidewalls of the superstructure, subsequently changing the width at the soffit. The top flange width, also the width of the bridge deck, is constant and the bottom flange width is changing along the bridge, widening at pier locations. This enlargement of the flange width creates a larger area for the load to spread, creating the fascinating "skirt" effect which adds a dynamic effect to the wide smooth soffit.

The variables, mainly the structural depth and the flange thickness, would be made in response to the bending moment diagram of the deck. From Risorgimento, we know that the web thickness of the box girder also changes together with the thickness of the bottom flange. A similar design concept can be assumed to have been used for Opera Bridge. Even though all variations can be justified through structural design values, the change in the web thickness might have made the bridge works highly labour intensive during its construction.

Timber formwork marks are very obvious on concrete surfaces, although these do not catch the eye since the scale of the bridge occupies viewers for a long time before such details are captured.

The piers of the Opera Bridge are uniquely their own and also among other Nervi Bridge designs. Massive and stunted piers are formed as a whole without a separate crosshead. The shape changes smoothly from an oval–circular base to a rectangle at the top with much larger sections in between. The columns used at the ramp structures are also unique with their cross-shape bases with smoothly curved planes transforming into a circular section at the top with a gentle taper.

This bridge occupies a considerable space in the city because of its robust shape and size. Luckily, the vicinity of the structure is still relatively open since no other large structures are built nearby. Fortunately, or by intention, the bridge has not been widened; otherwise, this would be the end of its aesthetic value.

Even though some tastes would consider this bridge robust, oversized or bulky, the author appreciates its scale and extraordinary character.

The junction was designed together with the bridge as well as the stairs on all sides provided for the circulation of pedestrians (Figure 7.31).

The bearings can be considered to represent the technology of their time very precisely. They are a combination of roller and rocker bearings used at the piers and abutments.

The current condition of the bridge exhibits considerable surface defects. There are many spalls and delamination of concrete and most are the result of a lack of drainage. The water leakage on the superstructure and the piers can be spotted from stains also resulting in some degree of rust developed in the bearings.

Efflorescence can also be observed in many locations, especially where there have been some utilities attached. The spalls, delamination, rusting and efflorescence make the structure a worthwhile site visit for engineers.

The expansion joints of the bridge are all filled with asphalt but still working despite this condition as seen from the cracks in the pavement (Figure 7.32).

Figure 7.31 Opera Bridge ramp and column detail on western end (author).

Figure 7.32 Opera Bridge details; left: pier, right: pier cap (author).

After visiting this bridge, the author recalled that this bridge site was a bazaar while the author was a student in Ankara. It was called "the Opera Bazaar" underneath the bridge from the 1980s until 2008. The bazaar was established by the municipality as a place to be dedicated for shop owners who were unable to access their shops as a result of conservation and development works in the historical centre of Ulus, which is 100 m away from the bridge. The municipality decided to utilize the ample spaces around the bridge for this purpose and constructed commercial pavilions for the shop owners.[31]

The bridge also presents a good case study in terms of utilizing the spaces underneath and near bridges. Some structures get very good treatments for these spaces, especially if there is no traffic under the bridge, creating an amazing environment for kids and communities.

Over time, the bazaar's designated spaces and kiosks had expanded and even diffused into the traffic lanes. Then, in 2008, all the bazaars were removed. Currently, these spaces are empty and unmaintained, housing street dogs and a disturbing smell. The current environment under the bridge is far from the initial purpose of being peaceful or attractive.

This unique bridge with its impressive features and contribution from a great engineer still serves its primary function of carrying vehicles in the heart of the capital, Ankara.

NOTES

1. Troyano, L.F. (2003). *Bridge Engineering: A Global Perspective*. Thomas Telford. UK. Page: 349.
2. The term used for Public Works was "imar", means civilization and development excluding the social content. Term later changed to "Nafia", which can be translated as utility, to refer to public assembly at the governance level. Nafia also later changed to "Bayındırlık", means prosperous, in 1935.
3. Province Album (1929) *Vilayet–i Hususi İdareleri Faaliyetlerinden*. Hilal ve Cumhuriyet Matbaaları. İstanbul.
4. Nafia (1933) *On Senede Türkiye Nafiası 1923 – 1933*. Nafia Vekaleti Neşriyatı. İstanbul. Page: 56.
5. Album (1929) *Vilayet-i Hususi İdareleri Faaliyetlerinden*. Hilal ve Cumhuriyet Matbaaları. İstanbul.
6. Municipality web site: https://www.mustafakemalpasa.bel.tr/ilcemiz/tarihce.html
7. Nafia (1933) Page: 71.
8. Nafia (1933) Page: 65.
9. Quataert, D. (1993) *Ottoman Manufacturing in the Age of the Industrial Revolution*. Cambridge University Press. UK. Page: 58.
10. Album (1929) *Vilayet–i Hususi İdareleri Faaliyetlerinden*. Hilal ve Cumhuriyet Matbaaları. İstanbul.
11. Nafia (1933) Page: 65.
12. Nafia (1933) Page: 71.
13. Petition of a contractor for document of certificate of proficiency, 1930. Source: Unclassified documents from the State National Archives–Republican Archives, KGM Fund, Binder no: 2020. Found in: Örmecioğlu, H. T. (2010) Technology, Engineering, and Modernity in Turkey: The Case of Road Bridges between 1850 and 1960. Ph.D. thesis submitted to Middle East Technical University (METU). Appendix. Figure H.3.
14. **Süreyya İlmen** (1874–1955) was the distinguished figure of military, commercial, political and cultural realm in the late Ottoman and early Republican period.
15. Küçüksu Köprüsü'nün inşaat albümü – Construction album of Küçüksu Bridge (1928-06-04) SALT Araştırma – SALT Research reached from: https://archives.saltresearch.org/handle/123456789/118983.
16. Figure 4.36 Up: Facade proposal for the Etlik Bridge with minaret-like newels. Down: Longitudinal section drawing of the Etlik Bridge, 1936. Source: Unclassified documents from the State National Archives–Republican Archives, KGM Fund, Binder no: 1839. Found in: Örmecioğlu (2010) Page: 155.
17. İlter, İ. (1989) Mimarlığımızın Neo-Klasik Donemi ve Bu Anlayışın Köprülerdeki Tek Örneği: Etlik Köprüsü. Karayolları Vakfı Dergisi. February 1989. Pages: 6–9.
18. The photographs during the earthworks of Municipality can be seen at SALT archives: 1955 Ankara imar planı doğrultusunda yapılan çalışmalar - Proceedings towards 1955 Ankara construction plan reached from https://archives.saltresearch.org/handle/123456789/128507.
19. Khorasan, an antique building material, used as a binder, developed in Western Asia, the Middle East and Anatolia. The word Khorasan, district to the east of Iran, is the name given to crushed and ground burnt clay as in bricks and roof tiles.
    Akman, S. et al. (1986) The history and properties of Khorasan mortar and concrete. 11. International Science and Technology Congress. İstanbul.
20. **Kemal Hayırlıoğlu** (1891–1984) was Nafia Chief Engineer for 15 years between 1928 and 1943. He worked in various positions in Nafia until retired in 1955. See Chapter 3 for expanded biography.

21. Hayırlıoğlu, K. (1934 September) Memleketimizde Köprücülüğe Bir Bakış: Yeni şose Köprülerimiz. Nafia Works Magazine. (Technical Part). Year:1/1. Page: 1.

22. İstanbul Metropolitan Municipality Taksim Atatürk Library (IBB Atatürk Library) collection: (Krt_017071) Diyarbakır'da bir köprü inşaatı.
    The footnote as: 'Erzan Köprüsü insaati: 24 Tesrin-i sani Sene 1339. Kasaba cihetinden alınmıştır. Boz derzler doldurulurken.'

23. **Feyzi Akkaya** (1907–2004) a prominent bridge engineer of the early Republican era. Together with **Sezai Türkeş** founded STFA Company, which was a leading engineering company in Turkey and worldwide. Their inventions and novel applications have been a vital contribution to the engineering field in Turkey. See Chapter 3 for expanded biography.

24. Akkaya, F. (1989) *Ömrümüzün Kilometre Taşları: STFA'nın Hikayesi*. Bilimsel ve Teknik Yayınları Çeviri Vakfı. İstanbul. Page: 93.

25. For the metal bridges designed with screw pile: see Chapter 06 of this book.

26. Nafia (1933) provides the total cost for Güzelhisar Bridge as 57450 TL with 121 m total length, which corresponds to 474 TL/m, unit cost.
    The next lowest cost reinforced concrete bridge is Adagide (BWS-01) bridge 64073 TL with 89.5 m total length, corresponds to 715 TL/m. The bridge similar in length and structural system is Çarşamba Bridge is 252964 TL with 274 m total length, corresponds to 923 TL/m.

27. Örmecioğlu, H. and Er Akan, A. (2012). An Unknown Work of Nervi: The Opera Road Bridge. Cantiere Nervi La Costruzione di Un Identità Storie Geografie. Skira Addison-Wesley Italia Editoriale s.r.l., Editör:Gloria Bianchino, Dario Costi.

28. Decision of the City Council of Ankara on commissioning Studio Nervi for a bridge construction on Opera Junction, 16 November 1968, no. 82 and Contract signed by Ekrem Barlas the Major of Ankara and Pier Luigi Nervi, 12 April 1969, The Archives of the Greater Municipality of Ankara. From: Örmecioğlu, H. and Er Akan, A. (2012).

29. BCA- 30-18-1-2_239_62_19_3 Letter for Nervi to be appointed to work for the Municipality of Ankara from 20.061969 to 20.06.1970 signed by President Cevdet Sunay.

30. Desideri, P., Nervi Jr., P. L. and Positano, G. (1979) *Pier Luigi Nervi*. Zanichelli Editore. Bologna. Italy. Pages: 144–147.
    Cresciani, M. (2007) Pier Luigi Nervi, bridge designer. In *IASS Symposium 2007: Structural Architecture - Towards the Future looking to the Past. 3–6 December 2007*. Venice, Italy.

31. Özkan, S.O. (2010) The (Re)Production of Public Spaces: A Survey on three cases from Ankara. M.Sc. Thesis Doctorate thesis submitted to Middle East Technical University (METU).

Chapter 8

Conclusion

This book introduced bridges constructed between 1923 and 1950, during which time the nation was rebuilding under challenging conditions. Nearly all these bridges can be regarded as heritage structures since they represent a certain period in the history of Turkey.

At the same time, the bridges themselves possess significant features as engineering structures. The aforesaid period also witnesses the worldwide spread of the evolution of stronger concrete with reinforcement after WWI, although steel production was still poorly developed in Turkey. Bridge building was being carried out with new materials, methods and technology in parallel with the industrial transformation of the country. Therefore, these bridges were also first reinforced concrete bridge, first suspension bridge, first cantilever bridge, first bowstring bridge and so on, constructed with the local engineering skills and resources.

Railways were forerunners of the industrial transformation in Turkey. The rail projects were generally constructed under better conditions being a part of major projects, with large companies attracted by a secure and quick financial return. The bridges for railways which had to carry much heavier loads, required special features for dynamic effects and had to provide uninterrupted service, all needed more demanding designs.

In the first years of the Republic, 1923–1932, engineering was still in its technological infancy. New bridge types and innovative methods were applied; however, the scale of engineering was still growing. The transportation infrastructure was constructed under basic conditions; nearly all construction labour was carried out by hand; concrete mixers and water pumps were an exception to the norm on the construction site (Figure 8.1).

While the country was still tending its wounds, the worldwide financial crisis in 1929 struck, leading to shortages, especially for concrete reinforcement, cement and steel. As a response, building authorities merged and management centralized, which turned out to be an effective policy. The situation also promoted self-sustainability in the country.

The 1932 Kömürhan Bridge was a technical marvel of its time in Turkey. Although designed and constructed by a foreign company, it was a considerable achievement to cross the formidable Fırat River with a permanent bridge by modern engineering and a demonstration of government service to an ignored part of the country. Kömürhan Bridge can be considered as a turning point in the evolution of engineering in Turkey. It was built by railway contractors engaged in the nearby construction. The bridge,

DOI: 10.1201/9781003175278-8

Figure 8.1 1935 Menemen Bridge abutment construction: dewatering of the groundwater and preparation of the gravel mix for foundations will be compacted with a hand hammer. Engineers can be noticed with their looks and hats. (SALT research, photograph and postcard archive.)

together with the ongoing railway projects, played a key role in elevating engineering standards and methods, and was an opportunity for local engineers to gain experience.

Kömürhan was followed by bridges like 1932 Silahtarağa and Sünnet, 1933 Aksu, 1934 Körkün, 1934 Pasur, 1935 Meşebükü, 1937 Keban Madeni, and later 1939 Pertek and 1939 Gülüşkür. All had noteworthy spans and designs, constructed by local companies. Especially, Pertek Bridge crossing over Murat, a branch of the Fırat, and again in the eastern part of the country, was built entirely by a local firm.

Engineering progress from 1923 to the 1940s continued evolving even though the 1929 world crisis and WWII resulted in material shortages and unstable conditions. The Second World War seriously affected Turkey, even though not involved directly in the war, because of its strategic location.

By the year 1940, construction technics, methods and conditions were developed and all contemporary road bridge types were constructed locally.

In 1947, Turkey received an aid programme from the US under the Marshall programme, existing roads were to be repaired and new roads were to be constructed. Nafia, Public Works Authority, transformed into a Highways Organization, which established formally in 1950 under the name General Directorate of Highways (KGM).

Bridge building after 1950 changed in nature with machinery and organized production, the design live loads increased and mechanical construction was taking place. The quality was not compromised; however, quantity production ruled the construction environment. Beam and Gerber bridges were built with standard designs, Arch bridges were still constructed, especially in the eastern regions, and steel bridges were still a solution for remote regions. The first prestressed bridge, İnandık Bridge in

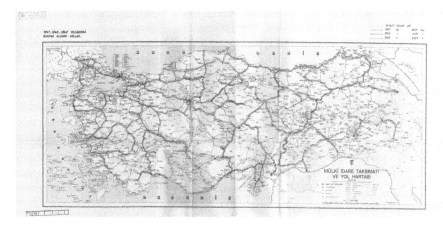

Figure 8.2 1949 dated map to show the roads repaired during 1948–1950 (BCA-30-10-0-0/156-95-24).

Çankırı, was built in 1963. Especially with prestressing, beam technology progressed to cross average spans more easily and the bridge domain changed its character with the beam bridges dominating (Figure 8.2).

Bridge building is a reflection of its time as a product of technology. At the same time, these endeavours depend on the social and cultural aspects, whose contribution depends on the size and strategic value of the bridge.

For early republican bridges, the social component is observed in multiple aspects. Firstly, they were the outcomes of a new modern regime that needed to demonstrate itself nationally and internationally. The authority took great care and control over the projects. Bridges were showcase for the government to prove their service, their modernity and technological improvements. The opening ceremonies and naming of bridges can be observed as evidence of their importance to society.

Second was the professionals at the backstage for the design and construction of the bridges. There was a common spirit with a fresh mindset who worked with dedication and a vision beyond their profession. Engineers of this era were exceptional characters who survived the war, probably migrated from their home within the borders of the empire, went abroad for education, were highly esteemed in society and were encouraged to use initiative. They were proud of the victory of serving their homeland and overcoming impossible situations.

These same engineers were contributing to the institutional setup of the construction industry in the country. For example, the US aid programmes were able to be utilized to form state organizations, like KGM, since the necessary professional environment existed. The same contractors also spread internationally in construction industry and gained reputation through their work.

The engineers were brave and dedicated, new types of bridges were designed and built, trial designs were carried out and innovative solutions were applied to the designs. Türkeş and Akkaya are good examples of representing this spirit and transitions; like most of their teachers, they did not even know the Latin alphabet at the beginning of their engineering education.

Their attitude was certainly shaped from the revolutionist, rational and diligent mindset of the Republic. Their dedication to profession, undisputable abeyance to ethics and prioritization of the quality of work over profits were also highlights of their era. The leader of the country, Atatürk, was a "hands-on" person, hardworking, intimate, caring and very determined character. He brought the new alphabet to the schools, hence given the title of "The Head Teacher".

The cultural component is more elusive compared to other parameters, as culture establishes over longer timespans. For the very first time, local engineers were engaged under the management of foreign companies, like the Hedjaz railway project. This futurist thinking played an important role in setting the cultural and mental background. The interaction and communication also stimulate the culture of learning from others, comparing and taking feedback for personal involvement through others and teamwork. All contributed to the culture with which the engineering profession evolved.

This book narrates their story through their work, although their devotion and success deserve specific appreciation.

Many of the early republican bridges are still in use after many years of service and nearly 3/4 of the bridges built during 1923–1940 still stand. Many of them did not cope with the increased traffic loads and new requirements, and are therefore abandoned often standing alongside the new bridges. A considerable number were lost due to flooding in the lake of dams built on the rivers. A few collapsed due to flood or were intentionally demolished to meet the demand of the new road networks. Many of the bridges are registered as heritage structures and are well preserved by the local and national authorities. KGM does regular work on preserving these bridges.

Republican publications, especially those for the 10th- and 15th-year celebrations, made critical comparisons with the Ottoman Empire, emphasising the greater number and modernity of the infrastructure built by the new Republic. Nowadays the early republican products are also ignored in the same way and underestimated with the glowing architecture of the empire. Modernization and progress should cooperate with history to learn rather than competing with it. The heritage, especially in this sense, is not a competition and comparison; it should all be about embracing what belongs to us and makes us.

Even though the bridges are physical structures, their preservation starts in the minds of the people: locals, users, professionals, carers, visitors, etc. The more the edifice is known, the more it will be recognized and cared for.

The writing of this book was very challenging for an engineering author since both the period and the strategic location of Turkey involved a sizeable amount of historical and political events. The original inspiration for this book came from the scarcity of studies in this area, hoping that it inspires a broader recognition of these bridges, their period and those involved in developing them.

Index